Epistemology Modalized

There are three primary aims of the book. The first, set out in the book's introduction, is to explain how two fairly recent developments in philosophy, externalism and modalism, provide the basis for a promising account of knowledge – an account that achieves anti-skeptical results and avoids Gettier-style counterexamples that are based on an agent having warranted beliefs that are merely luckily true. Epistemological externalism is the thesis that not all the factors that make a true belief a case of knowledge need to be cognitively accessible or "internal" to the agent. Modalism, as the author construes it, is the use of possible worlds talk in clarifying what would or might be the case and what is necessarily so.

The second objective is to work through the different modalized epistemologies extant in the literature, assessing their strengths and weaknesses, with a view toward constructing a working theory. Finally, the author proposes the theory that knowledge is reliably formed, sensitive true belief, and defends the theory against objections.

Kelly Becker is Assistant Professor of Philosophy at the University of New Mexico.

Routledge Studies in Contemporary Philosophy

1 Email and Ethics
Style and ethical relations in computer-mediated communication
Emma Rooksby

2 Internalism and Epistemology
The architecture of reason
Timothy McGrew and Lydia McGrew

3 Einstein, Relativity and Absolute Simultaneity
Edited by William Lane Craig and Quentin Smith

4 Epistemology Modalized
Kelly Becker

5 Truth and Speech Acts
Edited by Dirk Greimann & Geo Siegwart

Epistemology Modalized

Kelly Becker

Routledge
Taylor & Francis Group
NEW YORK AND LONDON

First published 2007
by Routledge
711 Third Ave, New York, NY 10017

Simultaneously published in the UK
by Routledge
2 Park Square, Milton Park, Abingdon, Oxon OX14 4RN

Routledge is an imprint of the Taylor & Francis Group, an informa business

First issued in paperback 2012

Typeset in Times New Roman by
Taylor & Francis Books

Library of Congress Cataloging in Publication Data
A catalog record for this book has been requested

British Library Cataloguing in Publication Data
A catalogue record for this book is available from the British Library

ISBN 978-0-415-95611-6 (hbk)
ISBN 978-0-415-54127-5 (pbk)

For Debra

Contents

Preface and acknowledgements ix

1 Introduction: Externalism and modalism 1

A. Externalism 5
B. Modalism 9
C. What should the theory do? 12
D. What's missing? 18

2 Process reliabilism 21

A. Goldman's causal theory 22
B. Goldman's discrimination requirement and relevant alternatives 24
C. Process reliabilism and why it is not enough 32
D. Implications for skepticism 38
E. Conclusion 38

3 Sensitivity 40

A. Nozick's subjunctive conditional theory of knowledge 41
B. Methods: An important refinement 47
C. Objections to Nozick's theory 52
D. Conclusion 60

4 Safety 63

A. Motivating safety 63
B. Weak and strong safety; luck and induction 65
C. Is safety necessary for knowledge? 68
D. Luck revisited: Safety requires a process reliability condition 69
E. Is reliability compatible with knowledge of the denials of skeptical hypotheses? 77
F. Conclusion 86

5 Knowledge: Reliably formed sensitive true belief 88

A. The theory 88
B. Problems and clarifications 96
C. Conclusion 109

6 Closure and the value problem 111

A. Closure 113
B. The value problem 128
C. Conclusion 137

Notes 139
Bibliography 165
Index of cases 171
Subject and name index 172

Preface and acknowledgements

I didn't see this book coming. When I left graduate school in 1999, I was focused on issues in the theory of meaning, having just defended a dissertation on meaning holism – a view I no longer accept. In the fall of 2001, Sandy Goldberg at the University of Kentucky invited me to drive up from Centre College (in Danville, Kentucky) once a week to participate in a reading group on DeRose and Warfield's anthology, *Skepticism: A Contemporary Reader*. I chose to present the nuts and bolts of Robert Nozick's tracking theory of knowledge, on which I had previously written a paper as an auditor in John Dolan's graduate course in epistemology at the University of Minnesota (back in 1991, when I was suffering in limbo as a businessperson). My interest in epistemology lay more or less dormant in the intervening decade.

Sandy's reading group apparently sparked something in me. I began thinking seriously about externalist epistemologies, and I even tried to convince myself that Nozick's entire approach was wrong. I wasn't able to do the latter, even though I realized that most practicing epistemologists had long since abandoned the centerpiece of Nozick's theory – the claim that S knows that *p* only if, if *p* were false, S wouldn't believe it. A substantial part of the aim of this book is to show that, with fresh eyes and an attention to detail, this crucial part of Nozick's theory looks pretty good. It looks even better when coupled with process reliabilism, a theory most vigorously developed by Alvin Goldman. Why? Because in the post-Gettier literature, epistemologists have scrambled to construct theories that eliminate the epistemic luck at the heart of Gettier's counterexamples, without explicitly noting that it has two distinct sources. Sometimes one obtains lucky true beliefs through unreliable means, and sometimes one obtains lucky true beliefs because of some quirky feature of the actual world. Process reliabilism eliminates the former kind of luck, and Nozickean sensitivity eliminates the latter, or so I argue herein.

As I say, I didn't see this book coming. So in order for me to write it, lots of things had to go right, and I needed help. I am grateful to the University of New Mexico for giving me paid leave during the Spring of 2005, which is precisely when I most badly needed the time off to work on the manuscript.

I want to thank my UNM colleagues, especially Russell Goodman and Fred Schueler, for their encouraging words and practical advice throughout this process. I want to blame Iain Thomson, my long-time friend, for goading me to write the book in the first place, but since everything turned out okay, I guess I'll thank him instead.

A number of people have given me detailed comments on parts of the manuscript, and they have my sincerest gratitude. John Greco, whose book *Putting Skeptics in Their Place* ought to be read by everyone with *any* interest in epistemology, helped me with the introductory chapter, as did my colleague and friend, John Taber. Alvin Goldman's comments on the second chapter were very useful, as were Sandy Goldberg's on the third chapter. Duncan Pritchard read most of the manuscript, and, with his characteristic charm and enthusiasm, brought to light many areas that needed refinement and improvement. (Read his excellent book, *Epistemic Luck*, to see how he would have liked to see the revisions of my Chapter 4, in particular, come out.) To the extent that we disagree over matters epistemological, I find myself thinking: With enemies like Duncan, who needs friends? Wayne Riggs gave me sage advice on the final chapter, and, as you'll see, his work also provides the foundation for much of it. I want to give special thanks to Brian Cooney at Centre College – previously a colleague, always a friend. He read most of the manuscript, offering detailed suggestions and, more importantly, constant support. I am also very grateful to the editors and editorial staff at Routledge, especially Max Novick, Erica Wetter, Tony Bruce, and Paola Celli for their support, advice, and timely responsiveness to all my rookie queries. To all of these people: Thank you. (Except Iain. I reserve the right to blame you for suggesting that I write a book.)

Finally, and most importantly, I am eternally indebted to my wife, Debra Hammer, for patiently and encouragingly abiding my fixation.

Part of Chapter 3 appears as "Is counterfactual reliabilism compatible with higher-level knowledge" in *Dialectica* 60:1 (2006), 79–84. Chapter 4 includes material previously published in *Metaphilosophy* 37:5 (October 2006). I thank both *Dialectica* and *Metaphilosophy* for permission to use these materials here.

1 Introduction: Externalism and modalism

Recent developments in epistemology, and in philosophy more generally, provide a promising foundation for an answer to a very old question: What is knowledge? The question stymied Plato in the *Theaetetus*, from which the traditional tripartite analysis of knowledge as justified true belief derives. It received fresh attention when Edmund Gettier showed that the three conditions of the traditional analysis were not jointly sufficient for knowledge. Subsequent attempts to repair the analysis of knowledge aimed (1) to amend the notion of justification to avoid the Gettier problem; (2) to add a fourth condition, for instance that there are no defeaters to one's justification; or (3) to replace justification with some other condition that captures the requisite link between belief and truth constitutive of knowledge.[1] The uniqueness of the third strategy is not clearly defined because one could easily argue that, whatever the necessary link between belief and truth turns out to be, it just *is* justification. Nonetheless, I see myself as pursuing this approach because the very term "justification" is all too pregnant with associated notions that I believe are not essential to knowledge, and working toward an account that explicitly involves justification as a necessary condition can lead us away from a proper understanding of knowledge. (A specific instance of this problem arises in Chapter 2.) Unencumbered by the requirement to explicate "justification," we can inquire into the requisite belief-truth link constitutive of knowledge by testing proposals for that link against our intuitions concerning whether an agent actually knows in particular cases. If we find that a correct or, at least, working account of that link does not capture the traditional conception of justification, then so be it.[2]

Our topic, then, is propositional knowledge: knowledge that p for some arbitrary proposition p. I will not claim that all other forms of knowledge, for instance, knowledge by acquaintance, knowledge of one's own phenomenological states, and know-how, are reducible to propositional knowledge, and so do not intend to give an account of knowledge in general. This only slightly diminishes the importance of an account of propositional knowledge, since it is through sentences and the propositions they express that we think and talk about the world. It would be a significant advance in our understanding if we had a plausible theory of such knowledge.

Following in the wake of Gettier's counterexamples to the traditional analysis of knowledge was a spate of papers and books that used the methodology of testing specific cases of knowledge, or lack thereof, against intuitions. It was the heyday of the counterexample method, and, though many epistemologists still use that method, many others are by now either tired of sifting through examples or, worse yet, convinced that the entire approach is either bankrupt or an exercise in mere bookkeeping (to borrow a term from Jason Stanley).[3] I for one am not persuaded; application of the method is indispensable for our purposes. I would go so far as to say that, recently, some philosophers have failed to properly acknowledge the deep pull of some time-honored examples of skeptical hypotheses in order to achieve theoretical gains. For instance, much recent literature (especially on, but not limited to contextualism and its theoretical cousins that aim to account for ordinary uses of "knows" locutions) has as its basis some hand-wringing about a seemingly paradoxical triad deriving from fundamental, shared intuitions: (1) that I (where "I" stands for any subject) know many ordinary empirical propositions; (2) that I do not know that radical skeptical hypotheses, for instance that I am a brain-in-a-vat (BIV), are false. However, together these seem to violate the intuitively compelling principle: (3) that knowledge is closed under known entailment – that if I know that p and know that p entails q, then I know that q (provided that I draw the relevant inference). For example, if I know that I'm typing at my computer, and that this entails that I'm not a BIV, then (it seems) I know that I'm not a BIV, but I don't know that.

One way to have one's cake and eat it, too, is to construct a theory of knowledge that, at least in some instances, allows that one can know that radical skeptical hypotheses are false. (Contextualists and safety theorists argue this. I discuss contextualism briefly below (§D) and again in Chapter 6. Chapter 4 is devoted to a discussion of safety.) This also allows one to maintain the closure principle, which is usually highlighted as a huge theoretical benefit. But how on earth does one turn that trick? Given the current climate in epistemology, it may seem somewhat reactionary to insist that our inability to know that radical skeptical hypotheses are false is a founding intuition of the entire enterprise. But you don't really have to take my word for it. After all, as I say, most of the contemporary discussions begin with the deeply held intuitive judgments stated above, and then, often, a theorist finds some clever way to undermine what is historically the most compelling and worrisome one, resulting in the claim that we can know that skeptical hypotheses are false. I may be charged with being obsessed about bookkeeping – with keeping track of our intuitions regarding basic cases – but at least nobody can accuse me of cooking the books![4] Our theory must be constructed on the basis of firm and lasting judgments about what we know and what we don't know, and only then can we assess the significance of any counter-intuitive theoretical implications, such as the denial of closure, as I do in Chapter 6. Working backwards, that is, from the desire to avoid awkward implications, is fine, but not if achievement of theoretical benefits requires us to revise the firm and enduring judgments which form the basis of our investigation.

It does not follow from the preceding that I am oblivious to legitimate misgivings of philosophers about the method of appealing to examples to elicit intuitive judgments about knowledge. In addition, I should note, too, that I am bothered, as a writer, by the fact that the method can cramp one's style. I aim herein to avoid some of the most egregious troubles thrown up by extensive use of cases and counterexamples. First, throughout the book, whenever I discuss a particular case, I give it a name in all upper case letters. In addition, I have included an index of case names with the page number where each first appears. I do this because I know from my own research how tedious it can be to try to keep track of all the different counterexamples to theories. Second, I do not insist that a plausible theory must pass every possible counterexample test, no matter how baroque. The fear that the approach is bankrupt seems to be that, once one has a recipe for generating counterexamples, no theory of knowledge can stand up to all of them. That may very well be, but surely some theories spawn more, and more obvious, counterexamples than others. While no theory may be wholly immune to them, we can and should hope that some theory will yield only counterexamples that are at best contentious. That would be an indication that we are close to hitting our mark.

Another frequently cited problem with appeal to intuition is that intuitions are untrustworthy; because they are informed by many of one's other beliefs, they are both unstable (because beliefs change) and somewhat subjective.[5] I actually think this is true, but it is also remarkable that, in the epistemology literature, very many of the examples engender almost unanimous verdicts. Again, we have to treat these as truth telling. Some cases, on the other hand, tug different people's intuitions in opposite directions. In clear cases of unclear verdicts, we ought to be ready to hand with possible explanations on both sides – why one might plausibly think that S does know that *p*, or why one could reasonably think S does not. I shall offer an analysis of knowledge that is stated abstractly enough that it admits of some leeway in *interpreting* the proposed belief-truth link, where the interpretations deliver opposing judgments in some very specific cases. When there is intuitional conflict, I leave it to the reader to interpret my proposed conditions on the belief-truth link, provided, of course, that she does so consistently. Alternatively, there may be a hidden ambiguity, for instance in cases of induction, about the content of one's knowledge: Is the relevant question whether one knows that a particular event will occur, or only that it is *highly likely* that it will? If it is the former, our analysis might yield a verdict of no knowledge, but if it is the latter, we may conclude that the agent knows. Cases of unclear verdicts arise more than once in this book, and I show how to accommodate both sides when necessary.

Throughout the rest of this introduction, I will set the stage for the subsequent proceedings. Before that, though, I want to lay my cards on the table. The theory of knowledge I will outline and defend holds that *knowledge is sensitive true belief that is formed by a reliable process.* S's belief is sensitive if and only if, if the proposition believed were false, S would not believe it. Sensitivity is reintroduced in Chapter 2 and treated at length in Chapter 3; a key issue will be the proper understanding of the sensitivity counterfactual. (The sensitivity condition in particular admits of some flexibility in interpretation, in that there may be more than one specific plausible reading of "if *p* were false, S would not believe that *p*.") I should note that sensitivity has fallen on hard times recently. Some of the most serious and influential criticisms of it hold that it is too restrictive – that the condition is violated in genuine cases of knowledge, including higher-level knowledge. I reply to these criticisms in Chapter 3 in the midst of my larger project of resuscitating sensitivity as a necessary condition for knowledge.[6] Sensitivity is no longer fashionable, but I think it is a lot more durable than its detractors have claimed. In addition, when process reliabilism is coupled with sensitivity, a view that I shall motivate in various ways throughout Chapters 2–5, the famous generality problem for reliabilism admits of a neat solution. Thus I am advertising a theory that both responds correctly to our basic judgments about specific cases *and* yields significant theoretical benefits.

Generically speaking, a reliable process is one that produces mostly true beliefs. However, I shall argue in Chapter 2 that a process is truly reliable only if it produces mostly true beliefs throughout nearby possible worlds (which include the actual world). That is enough to indicate the basic contours of the theory I favor.

Also in the spirit of revealing my hand, I want to make it clear that the close philosophical precedents for the theory articulated and defended herein are not at all lost on me. The pioneering works of Alvin Goldman and Robert Nozick are the main influences, and each gets extensive treatment in this book (in Chapters 2 and 3, respectively). Indeed, depending on the particular writings of Goldman upon which one focuses, one could come to see the present work as a defense of a version of his theory of knowledge, give or take. But Goldman is far more interested than I am in explicating some notion answering to the requirements of traditional justification, and this has led him to change his view of the modal aspects of knowledge over time and to become involved in issues that I think only muddy the waters. In addition, it is not entirely clear to what extent he accepts and applies the sensitivity condition, though it does appear famously in his early discussion of perceptual knowledge. Finally, since the days when reliabilism was essentially the only externalist theory around, new problems for it have emerged, as well as new interpretations of old problems. In setting out as clearly as I can the requisite reliability and sensitivity conditions on knowledge, in resuscitating sensitivity as a crucial ingredient in an account of knowledge, in explaining the complementary nature of reliability and sensitivity, and in showing how problems that afflict reliability and sensitivity severally can be solved by integrating them, I aim to extend and provoke discussion of modalized epistemology.

I have set out a preliminary sketch of the theory advanced herein for two reasons: first, to give the reader a sneak preview of my conclusions; second, to introduce the important recent developments alluded to in the very first sentence of this introduction. What has happened in recent philosophy that makes possible a satisfying resolution of our investigation into knowledge?

A. Externalism

In the second half of the twentieth century, a kind of paradigm shift led many philosophers away from some deeply entrenched ideas about the *location* and *epistemic accessibility* of features constitutive of phenomena about which we theorize. Though of course our topic in this section is *epistemological* externalism, I want to digress briefly to describe a parallel movement in the theory of meaning. There, it was assumed, for some very good reasons, that the meaning of an expression must somehow supervene on states of the individual who thinks or utters it. One very good reason for this assumption is that a theory of meaning is part of a general theory of propositional attitudes, and the latter is part of the theory of action. A propositional attitude, for instance a belief, is individuated in part by its content, or meaning. In turn, how one acts depends on (at least implicit) practical reasoning involving propositional attitudes. It is because Smith believes that there is water at the drinking fountain and desires that she have a drink of water that she walks over to the fountain and takes a drink. This kind of action explanation presupposes that Smith knows what she believes and desires. Smith might be wrong that there is water at the fountain, but surely she knows, in some direct and privileged way, that she *believes* that there is water at the fountain when she believes it. However, if Smith's attitude contents are individuated at least in part by features of the environment that are *external* to her, this threatens to undermine her privileged kind of knowledge about her own mental states, because knowing which mental state she is in would seem to require knowledge of her environment, which appears to be a more substantive achievement than knowledge of one's own mental contents. That threat to self-knowledge, in turn, puts a deep wrinkle in the theory of action predicated on privileged knowledge of one's propositional attitudes.

Be that as it may,[7] externalism in the theory of meaning is now pretty much the orthodox position. Hilary Putnam's twin earth thought experiment shows that two "twins" who are molecule for molecule identical, but whose environments differ only in the chemical composition of what they each respectively call "water," refer to different things with that expression, hence the truth conditions of their respective "water" utterances and thought tokens are distinct.[8] (That is, semantic externalism is committed to saying that the twins' expressions do *not* share referents – for instance, "water" does *not* refer to just anything that has the same phenomenal qualities as the stuffs each twin, respectively, calls "water.") Stipulating that the twins are identical means that they have all the same (non-semantic) dispositions, from which it follows that each uses her term "water" to talk about a colorless, odorless liquid that fills rivers and lakes, that one drinks and bathes in, and so on. Non-semantically speaking, they have the same *speech* dispositions. On the assumption that "water" is a natural kind term, it refers only to H_2O, hence any environment without H_2O is one without water. Were one to learn to use the expression "water" in such an environment, one would be referring to something other than water. It seems to follow that, unless an agent knows which kind of environment she is in, she does not know what her expression "water" refers to.

Tyler Burge (1979) extends anti-individualism, or externalism, beyond the treatment of natural kind terms (and names), and expands it explicitly from just the theory of reference to the theory of sense itself.[9] He argues that the concepts associated with many everyday terms, such as "arthritis," "sofa," and "brisket," get their correct application conditions partly from features external to the individual, and that such concepts are constituents of thoughts, or propositional attitudes. He contends that, when an agent is competent in the use of a term – when, by and large, she uses it correctly – she possesses the associated concept, even if she does not fully understand its application conditions and thus sometimes *misuses* the word (and not merely due to misperception and similar empirical errors). Burge argues forcefully that, in many cases, usage that conflicts with the best expert understanding of a term is not *correct*, idiosyncratic application of a term deriving from the speaker's idiolect, but is rather *incorrect* application of an expression whose correct application conditions are partly determined by experts. A weighty source of evidence for this claim is that ordinary speakers defer to experts and stand corrected when their usage goes astray, which implies that they take themselves to be beholden to external norms of correct application.

It is this last point that is especially compelling. A commonly taken upshot of W.V. Quine's indeterminacy of translation argument in *Word and Object* and some interpretations of Ludwig Wittgenstein's private language argument in *Philosophical Investigations*, including Saul Kripke's very influential reading, is that there is no fact about an individual, taken in isolation from her community, that determines the meanings of her words.[10] The central problem, in short, is that meaning is an inherently normative notion, in the sense that words have correct and incorrect applications. Skepticism about meaning, then, is sometimes skepticism about norms. All of my past usage of a term is consistent with any future usage under *some* rule. So which rule am I using? To what fact can I point to determine the rule? Externalism can answer these questions, or at least provide a key to answering them: The rule is determined by the actual nature of one's physical and social environments. We hereby purchase a plausible account of the location of norms constitutive of meaning at the potential cost of sacrificing privileged self-knowledge of propositional attitudes.

A similar shift occurred in epistemology. At least since Descartes, it has generally been assumed that one knows only if one has cognitive access to some ground that supports her belief. It seemed obvious that if the person is not in a position to back up her belief, the belief is not sufficiently tied to truth to count as knowledge. This is an internalist claim about justification, because it says that a belief is justified only by factors that are cognitively accessible to the agent and recognizable on reflection.[11] If the agent is not poised to produce justification, surely she does not know.

As in the case of meaning, there are very strong reasons for cleaving to the traditional view, though the reasons are less theoretical than intuitive. To think that one could know that *p* without knowing whether the justification conditions, satisfaction of which is necessary for knowing, are fulfilled just sounds odd. Perhaps this is because most of us are introduced to epistemological problems by reading Descartes. Though it is methodologically presumptuous to take one's own case as representative, I doubt that my trajectory through the study of epistemology is idiosyncratic, and so I would like to describe my original difficulty in seeing the relevance of anti-internalist conceptions of the belief-truth link to knowledge. When I first read Alvin Goldman's "A causal theory of knowing" as an undergraduate, I was convinced that Goldman was unwittingly talking about something other than knowledge when he postulated that knowledge is true belief standing in an appropriate causal relation with the fact that makes it true.[12] It *had* to be wrong as an account of knowledge, for it did not require that the agent knows that the causal relation in fact held. So how could she have any knowledge? Goldman's causal theory is a version of externalism, precisely because it does not require internalist justification, as (very) briefly sketched above.

My early and perhaps naïve misgivings about the options available to the epistemologist concerning the requisite link between belief and truth led me to think that the prospects for a theory of knowledge are bleak. Internalism seemed to lead directly to skepticism, for if all of my experiences are consistent with me being a BIV, then on what internally accessible grounds can I justify my belief that I am not a BIV? And if I do not know that, then how could I know, for example, that I am sitting here typing? On the other hand, externalism seemed out of the question because it had merely changed the subject from knowledge to something else, or so I had believed. I have since had a change of heart, or mind. A few years ago, in preparing a general course in epistemology, I had occasion to reread Goldman's seminal causal theory paper. In it is an example concerning knowledge that Abraham Lincoln was born in 1809. We can imagine that someone read about Lincoln's birth date long ago and now firmly holds that Lincoln was born in 1809, but has absolutely no recollection of where she obtained this fact. We can suppose, rather unfancifully, that the person does not remember reading or hearing it, and so can offer no grounds for her belief, even though she in fact read it in a reliable source. It strikes me as utterly overblown to say she does not know because she cannot produce justification. And it strikes me that lots of knowledge, including much everyday perceptual knowledge, is just like that.[13] I realized that externalism is not implausible.

It is important to note that externalism is *not* incompatible with instances of internalist justification. Externalism is simply the thesis that knowledge does not *require* that *all* justificatory or warranting factors are knowable upon reflection. Knowledge is true belief where there is an appropriate link between belief and truth, and that link may or may not be cognitively available to the agent. This is significant because it is sometimes suggested that internalism and externalism constitute a false dichotomy that needs to be overcome, that the correct theory of the belief-truth link constitutive of knowledge is a combination of the two.[14] There is no such thing as a combination. If you believe that internalist justification is not a requirement for knowledge, but that some condition beyond mere true belief is, you are very likely an externalist.

The proper understanding of the internalism-externalism debate in epistemology is a vexed issue. It is not important, for present purposes, that we characterize the issue in a way that accords with *everyone's* thinking about it, but it is crucial to have a clear working conception of the debate. To that end, let us defer to Matthias Steup's characterization of internalism, according to which justificatory factors must be "recognizable *on reflection*" (Steup 2003, 310). Thus for Steup "the *reliability* of cognitive processes does not qualify" as a justificatory factor. The implication is that all justificatory factors must be internally accessible to the agent.[15] Again, one may not accept this as the canonical version of internalism, but it has the virtue of being entirely clear. It follows that anyone who thinks that factors not cognitively accessible to an agent can justify (or warrant) that agent's beliefs is an externalist, even if one also thinks that there must be some cognitively accessible warranting factors if an agent's true belief is to constitute knowledge.[16]

Externalist about what? The belief-truth link. In most cases, excepting self-knowledge, the truth (say, that there is water at the fountain) is in some obvious sense external to the individual, so both internalism and externalism, as we understand them here, must accept *that* much externalism, but that is beside the point. And as we have seen, if belief contents are individuated externally, then there will be an externalist element in the belief condition on knowledge, too, whatever one's view of the belief-truth link.[17] So again, the relevant issue concerns whatever condition for knowledge it is that ties one's belief to the truth, traditionally construed as justification. I prefer the general term "warrant" for the condition that, together with true belief, constitutes knowledge, in large part because of my qualms, briefly described at the beginning of this introduction, about the preconceptions loaded into the notion of justification.[18]

Freeing ourselves from internalism opens avenues to an account of plausible cases of knowledge where internalist justification is absent. It also provides a basis for a reply to an extremely vexing skeptical argument, briefly mentioned above, that knowledge is impossible. I shall return to this point later in the introduction. Throughout this book, we will work through various externalist attempts to spell out the needed belief-truth link with the aim of arriving at the right one. Beware, then, that this is a book unapologetically ensconced in the externalist framework.

A conjecture of mine about the traditional dominance of internalism in philosophy is that philosophers are often, ironically, the least well suited to investigate some of their most cherished topics. Philosophers by nature are not interested in obtaining brute knowledge; they seek understanding. It is not surprising, then, that in their accounts of knowledge, they have customarily sought something more than what the relatively easily achievable positive epistemic status calls for. A philosopher's life is one of rational engagement, where the paradigm of knowledge is well-grounded (and trenchantly argued) true belief. However, not all knowledge fits that mold. In short, philosophers have tended to over-intellectualize their problem domains – to inflate the statuses both of our knowledge of our own mental contents and of our knowledge about our own knowledge – and they have paid the price for it, until recently.[19] My suggestion is that, in both semantics and epistemology, externalism is a corrective for this traditional over-intellectualization.

B. Modalism

The second main development paving the way for a successful theory of knowledge is the free use of possible worlds talk in explicating modalities like possibility, necessity, and what would or might be the case. A statement is necessarily true if true in all possible worlds, necessarily false if false in all possible worlds, and contingently true if true in the actual world but not in all possible worlds. (I ignore complications raised by physical necessity – truth in all worlds consistent with actual physical laws.) Kripke's seminal *Naming and Necessity* has been widely influential in this movement, that is, in the exploitation of possible worlds discourse for wider philosophical purposes.[20] I doubt that his work is a panacea for all one's worries about modal expressions like "would" (be the case), for whether we deem a statement true in this or that possible world depends on our intuitions, and those, as I have already said, can be unstable or conflicting. Still, thinking about what is the case in close possible worlds proves to be very useful for our inquiry.

Suppose we start with Goldman's (1967) original causal externalist account of knowledge. I see a bird and recognize it as a certain species *s*, and it being that species in fact causes me to believe it is, based on my perception of it. On the causal account, I know that it is an *s*. But suppose, too, that there is another species of bird that I cannot distinguish from *s*, and that species happens to live around here, too. It would seem that I do not know that this particular bird is an *s*, because there is a close possible world where the bird I look at is a member of the other species, and I believe it is an *s* anyway. Thus the following conditional is true: If it were not an *s*, I would believe it is. (This example conforms to the possible world's treatment of counterfactuals proposed by David Lewis.[21]) For our purposes, the truth of the conditional shows that knowledge requires more than that one's belief is appropriately tied to the truth in the actual world; it must be so tied in nearby worlds, too. Regardless of how we ultimately construe the need for belief to be true throughout a range of possible worlds, this general approach is what I mean by "epistemology modalized."

Some philosophers, including many of my colleagues, feel an intense aversion to possible worlds talk. In my view, there are better and worse reasons for the aversion. Kripke takes pains to forestall one of the potentially good reasons. If one takes the notion of a space of possible worlds literally, one gets the impression that a possible world is a place to which one could travel, say on an advanced spaceship. Lewis's view that possible worlds are no less real than the actual world sometimes seems to encourage this idea. (Lewis surely did not mean to encourage it himself. On his realist, possibilist view, non-actual possible worlds are spatiotemporally distinct from the actual world.) The weird ontological status of possible worlds leaves many philosophers with the queasy feeling that the universe is populated by very curious entities whose nature is not well understood. Kripke asks us to think of possible worlds talk as a useful (but perhaps misleading) way of talking about ways things might have been without committing us to the reality of possible worlds. Thus to say that there is a not very distant possible world where Nixon lost the 1968 election is simply a vivid way of saying that Nixon might have lost the election. Or to say that, in the closest possible worlds where I am a millionaire, I travel more often, is to say that, if I were a millionaire, I would travel more often. We can understand these claims – both the subjunctive and possible worlds versions – without presupposing that possible worlds are no less real than the actual world.[22]

Another reasonable worry about possible worlds talk is imprecision in our understanding of the truth conditions of the counterfactuals explicated by reference to them. The concern is that God only knows *precisely* how far one must "travel" from the actual world to determine which worlds are relevant to assessing counterfactual conditionals, allowing only God to know what exactly makes true our claims about what would be the case if things were different than they are.[23] Rather than casting doubt on the use of possible worlds discourse, however, this sort of concern motivates it. Take the following example. If I were a millionaire, maybe I would travel more, but then again, maybe I would become so decadent and lazy as to never leave the couch. It may be difficult to determine which "would" claim is true. And yet it is plausible to say that the second of the two counterfactuals makes reference to a more distant possible world than the first. I know that in the closest possible worlds where I am a millionaire, (1) I have more money than I do now; (2) some people ask me for money occasionally; (3) I invest at least some of it; (4) I give at least some to charity; (5) I eat well; and (6) I travel more. In worlds that are close to (or like) the actual world, except that I am a millionaire, propositions (1)–(6) are true. It is a more distant world where I *also* become decadent and lazy. (I am not now decadent and lazy, so unless becoming a millionaire directly, perhaps automatically, causes me to be decadent and lazy, the world where I am a millionaire, decadent, and lazy is intuitively more distant than one where I am a millionaire but my fundamental character does not change.) The point is, then, that while we may encounter difficulties in evaluating what would be true in some counterfactual situations because we do not fully understand which possible worlds are relevant to assessing counterfactual claims, we at least have a pretty strong, though admittedly not perfectly precise, intuitive understanding of the space of possible worlds.

In Chapter 3, for example, where the sensitivity condition is discussed at length, we will encounter some difficulty in interpreting the subjunctive conditional, if *p* were false, S would not believe it. It will be useful to think more specifically in terms of possible worlds. Indeed, this is where the interpretive leeway, briefly discussed at the beginning of this chapter, arises. Instead of laboring to find some canonical distance of the set of possible worlds relevant to assessing sensitivity, one may appeal to one's own intuitions about the possible worlds through which S's belief must be sensitive to the (truth and) falsity of *p* to constitute knowledge, and then rely on those intuitions to determine the preferred reading of the sensitivity conditional. Those intuitions will sometimes admit of interpersonal divergence, but at least the divergence will be clarified in terms of the disagreement about possible worlds, and thus we can get beyond a mere verbal dispute about the meaning of "if *p* were false, S would not believe it." I am sanguine about the prospect that this can legitimate our talk of possible worlds in a way that helps us find a plausible modalized theory of knowledge.[24]

I consider three core versions of modalism in this book, one each in Chapters 2–4. The first holds that knowledge is true belief formed by a reliable process, one that produces mostly true beliefs. There need be no modal dimension to this theory, but by the time we reach the end of Chapter 2, we will see that a process being truth-conducive only in the actual world is insufficient for bona fide reliability, hence a truly reliable process is one that is *modally* reliable – one that I think is best characterized by truth-conduciveness not only in the actual world but also in nearby possible worlds. Otherwise there is far too much room for epistemic luck – for the possibility that one's belief is true but merely luckily so. In fact, it is a central motivation of modalized epistemologies to eliminate this kind of epistemic luck. The notion of epistemic luck is thus very important, and I shall say more about it below (§C.2). The second theory (Chapter 3) holds that only sensitive true beliefs constitute knowledge, where, again, sensitivity is the modal condition which says that S knows that *p* only if, if *p* were false, S would not believe it. The third modal principle, highlighted in Chapter 4, holds that knowledge is true, safe belief, where a belief is safe if and only if, in nearby worlds where S holds that belief, it is true. In Chapter 5, I defend a theory that incorporates modalized process reliabilism and sensitivity but rejects safety.

It will be useful to settle some terminological matters at this early stage of our inquiry. Chapter 2 discusses process reliabilism, a term that I will use synonymously with global reliabilism, whereas Chapters 3 and 4 focus on versions of local reliabilism. We see already that a gloss on the notion of *process* reliabilism is in order because the term "reliabilism" has been attached to many distinct theory types. Taken by itself, the term does not distinguish between the reliability of a belief-forming process with respect to a particular belief (local reliabilism), wherein a process might be said to be reliable *in that instance* if the belief in question is true throughout a range of possible worlds, and the more general reliability or truth-conduciveness of a belief-forming process (global reliabilism). Following Goldman, let us say that a process is reliable if it produces a high ratio of true beliefs, such that "reliability consists in the tendency of a process to produce beliefs that are true rather than false."[25] Taking general truth-conduciveness as a basic feature of reliability, it is dubious whether local reliabilism is aptly named, insofar as it applies not to the truth propensity of the belief outputs of a process taken as a whole, but only to particular beliefs. But let us follow the common usage such that both the local and global versions are kinds of reliabilism.[26] While it is true that both local and global reliabilism make reference to processes, only the latter focuses on the overall truth propensity of the process, rather than on specific beliefs, which requires us to distinguish global, or process reliabilism, on the one hand, from local reliabilism on the other.

C. What should the theory do?

1. Be accurate

The section header may sound like a ridiculous question – the theory should tell us what knowledge is. But it really is a substantive question because some philosophers believe that a theory of knowledge should explicate the notion of justification, or that a theory must give a refutation of skepticism. I disagree. The main criterion for success is that the theory must respond to our considered judgments and intuitions about cases of knowledge; we can then unpack its consequences. Since almost everyone thinks that human beings are capable of possessing knowledge, the theory will likely yield some anti-skeptical consequences, but that does not mean that *refuting* skepticism, that is, demonstrating that we in fact possess knowledge, is itself an aim of the theory.

2. Avoid epistemic luck

The topic of epistemic luck has generated much discussion recently, one result of which is the distinction between (at least) two kinds of luck that are prima facie threatening to knowledge. Duncan Pritchard distinguishes veritic and reflective epistemic luck, where a belief is veritically lucky "if, and only if, the agent's belief that ϕ is true in the actual world but false in nearly all near-by possible worlds in which the belief is formed in the same manner as in the actual world."[27] If I believe truly that my friends bought me a guitar, but I believe it only because I wish it to be so, then my belief is veritically lucky. (This and similar examples suggest that Pritchard sets the bar for veritic luck a bit too high in claiming that a lucky belief is one that is false in nearly all nearby possible worlds. There may be many nearby worlds where my friends buy me a guitar, but there are also many, though not nearly all, where they do not, and that is enough to demonstrate the luck involved in true beliefs based on wishful thinking. Whether a belief must be true in many, most, or all nearby worlds to satisfy modal conditions on knowledge is a question we will consider throughout this book.) The basic idea is that a veritically lucky belief might easily have been false, and, as we have just seen, we can exploit possible worlds talk in clarifying "what might have been."

By contrast, a belief is reflectively lucky just in case it is true in the actual world but false in nearby worlds *consistent with what one knows by reflection alone.*[28] Pritchard (2003, §4) cites the frequently invoked chicken sexer as an example of reflective luck (CHICKEN SEXER). A chicken sexer is someone who can determine the sex of a young chick, but has no idea how. Though the chicken sexer's beliefs may be true in the actual world and in nearby worlds, in nearby worlds "consistent with what the agent is able to know by reflection alone" (Pritchard 2003, 122), her beliefs are false. There is nothing available to her by reflection that would tip her off, and so the chicken sexer's beliefs are reflectively lucky. The phenomenon of reflective luck suggests to some that there must be an internalist justificatory condition on knowledge – one that is typically responsibilist, involving an ability to fulfill one's epistemic duty in forming beliefs.[29] The crucial issue here is whether it accords with our intuitions to say that the chicken sexer knows, assuming that her beliefs are reflectively but not veritically lucky. Suppose, then, that there is some (unknown) reliable mechanism that allows the chicken sexer to determine the sex of a chick, and that this mechanism is "stable" enough that it would yield true beliefs throughout close possible worlds. Her belief is not veritically lucky, but it is reflectively lucky. It seems to me quite plausible that the chicken sexer knows, but of course, I am an externalist, so I would think that.[30] The point is that our main concern as externalists, whose primary aim is to identify the severally necessary and jointly sufficient conditions for knowledge, is veritic rather than reflective luck.

It is important that our aims are so understood. Pritchard (2005) argues that the inability to eliminate reflective luck is at the root of skeptical arguments aimed to show that we do not (rather than cannot) possess knowledge. If we cannot show the skeptic that our beliefs are not reflectively lucky, we cannot *demonstrate* that we actually have knowledge, but this is compatible with our having it nonetheless, provided that our beliefs are not vertically lucky. So it would be wrong to say that reflective luck is of no concern, but its elimination is not a necessary condition for knowing. I say more about the relevance of the distinct skeptical claims – that knowledge is not possible and that knowledge is not actual – in the next section.

Our most basic guiding presupposition is that a successful theory simply aims to give us a plausible explication of knowledge that coheres with our most stable intuitions. One of our deepest intuitions is that beliefs that are veritically lucky fall short of knowledge. The original Gettier cases involve double luck – good and bad (GETTIER). Suppose that Smith has very good evidence that Jones owns a Ford. On this basis, he comes to believe, as a logical consequence, that either Jones owns a Ford or Brown is in Barcelona. It turns out that Brown is in Barcelona, but that Jones does not own a Ford. Still, Smith's belief is well grounded (justified, warranted) and true, but he does not know.[31] His disjunctive true belief is veritically lucky, and its basis is unluckily false. Students first exposed to Gettier cases often think of them as simply involving some kind of logical trick that is psychologically implausible: "Why would Smith believe some disjunctive proposition based solely on evidence that one of its disjuncts is true?" But the "why" question is totally irrelevant. In fact, once you see the basic point, there is a clear recipe for generating Gettier cases: Imagine an agent whose good grounds for belief are unluckily false, upon which she forms a true belief that is veritically lucky – one that might easily have been false. (Schematically: S has good grounds for believing that *p*; believing that *p* provides good grounds for believing that *q*; based on S's belief that *p*, S infers that *q*. *p* is, in fact, false, but *q* is true.) A basic assumption at work here is that one can have a false warranted belief, one based on good grounds or, more generally and compatibly with the view I will defend, on a reliable belief-forming process. That a belief is reliably formed does not guarantee that it is true.

Zagzebski (1996, 283) argues that any account of warranted belief that does not entail truth is susceptible to the kind of double luck that generates Gettier cases because, so long as there is a gap between warrant and truth, there is room for bad epistemic luck. All we need to do is dream up a scenario where bad luck is coupled with some good luck such that the belief is true anyway, and we have a Gettier case. I think that Zagzebski is right, but interestingly, even on some accounts of the belief-truth link that entail that the belief is *true*, there is room for counterexamples that, while not exactly falling under the rubric of veritic luck, are very like Gettier cases. In typical Gettier cases, there is an element of good luck, without which the belief might easily have been false. Gettier-style examples, however, do not entail good luck (in this sense anyway), and therefore can be constructed for theories aimed to eliminate such luck. Take the sensitivity principle that I will advocate in addition to process reliability. Tweaking our original definition, we can say that sensitivity entails that a true belief is knowledge only if, in the closest world(s) where the proposition believed is false, the agent does not believe it. Well, the closest possible world is the actual world. If S in fact sensitively believes that a proposition is true, it cannot be false (for if it were, S would not believe it). Is sensitive true belief immune to Gettier-style counterexamples? Surely not. Suppose Jones intends to mislead me into forming a *true* belief on *false* grounds.[32] She only offers those grounds if the proposition is true. Given that my grounds for believing are false, I do not know, but it's not the case that my belief might easily have been false (in the sense captured by the phenomenon of veritic luck, anyway) – if it were false, I would not believe it because Jones would not have given me a reason to believe it. My belief is therefore not lucky according to the definition of veritic luck given above. There is, of course, a sense in which it is *unfortunate* that I have such a devious informant, but the belief is not veritically lucky. This kind of case does not undercut the need to eliminate veritic luck; rather, it shows that eliminating veritic luck by itself is not enough to steer clear of plausible counterexamples to our theory of the belief-truth link.

We need to capture the sense in which a true belief that is not veritically lucky may still fall short of knowledge even on externalist grounds. In the Jones case, I claimed that it is unfortunate that I have a devious informant. So bad luck seems to be relevant, but veritic luck does not. The problem is that the process by which I form the belief is not sufficiently reliable. This diagnosis requires us to note two things: first, that veritic luck is defined by reference to individual beliefs and the specific way each is formed; second, that there must be a distinction between the general processes by which one forms beliefs, and the specific way or method by which one forms a particular belief. In Chapter 5 I offer criteria for distinguishing methods from processes, but we need not bother with the details here. For now, I want to say that knowledge is incompatible not only with veritic luck, but also with what I will call *process luck*:[33] A belief that ϕ is formed by a lucky process if, and only if, the agent's belief that ϕ, formed by the method actually used, is true in the actual world and in nearby worlds, but the process by which it is formed produces a significant ratio[34] of false beliefs in many nearby possible worlds. The idea is that, even though one's belief might not easily have been false given the *method* (or manner, to use Pritchard's term) by which it was formed, beliefs formed by the more general *process* employed tend to be false, and so it is lucky that the particular belief in question turned out true, given that process. If the root worry about veritic luck is that a belief might easily have been false, then we must also attend to process luck. Suppose that, in the Jones case, the method is something like: If Jones gives me strong evidence that *p* and *p* entails *q*, believe that *q*. Again, since Jones only offers evidence that *p* provided that *q*, my belief is true throughout nearby worlds, hence not veritically lucky. The belief-forming *process*, on the other hand, for example, basing belief on Jones's testimony, is more general than the method. (In Chapter 5 I will argue that individuating processes so narrowly as to include, for example, a specific person's testimony, is well motivated, hence not ad hoc, but nonetheless allows for a neat distinction between methods and processes.) On the plausible assumption that Jones is prone to mislead, it makes sense to say that the process would produce all-too-many false beliefs, and thus that my belief is formed by a lucky process. It is a matter of substantial luck that this particular belief is true, given the general unreliability of the process.

By the time we have constructed our theory in Chapter 5, one that includes both process reliability (anti-process luck) and sensitivity (anti-veritic luck) conditions, we will see that it is immune to *almost* every kind of Gettier case. I do not doubt that some such example could be constructed against our theory, but I doubt that the case will be all that compelling. It is one thing to imagine that one is a BIV or the victim of an evil demon – exotic but fairly simple scenarios that provoke us to think seriously about the requirements for possessing knowledge of even the most simple kinds.[35] It is another to construct very convoluted cases to achieve a Gettier example that afflicts a theory with both a sensitivity component and a process reliabilism component. The more convoluted the case, the less worrisome.

3. Explain how knowledge is possible

One would almost have to be a professional philosopher to take seriously the idea that nobody can ever know anything, and yet such skeptical arguments have exerted enormous influence in epistemology. These two points imply that our theory must at once explain how knowledge is possible and be sensitive to the power of skepticism. There are, of course, many different kinds of skepticism, and though I won't try to delineate all of them, there is one distinction that is especially salient for us – between skeptical arguments that aim to show that knowledge is *impossible* and those that try to show that *nobody in fact possesses* any knowledge. Since the former is stronger, it ought to be easier (though by no means easy) to answer it, but failure to distinguish the two can lead to confusion over what exactly is to be sought in a theory of knowledge.

Christopher Hill unpacks a version of the traditional skeptical argument that knowledge is impossible.[36] The main move is to introduce a skeptical scenario that is consistent with all of one's experience, say, that I am a BIV. Because all of my experience supports the hypothesis that the world is as I believe it to be no more than it supports the BIV hypothesis, I am not justified in believing the former; indeed, I *cannot* achieve such justification, and therefore *cannot* know that the world is as I believe it to be. Externalism generally, and in particular the theory I shall argue for in this book, provides a reply to this argument. Briefly, I *can* know ordinary truths about the world if my beliefs are in fact reliably formed, whether I know them to be or not – whether I know that I am in a world that is compatible with the reliability of my cognitive processes or not. Knowledge need not require the kind of justification demanded by the skeptic.

Ernest Sosa argues, in fact, that this approach can give us a fully general theory of knowledge – one that applies to all cases of propositional knowledge, including knowledge that the theory is true.[37] So long as certain conditions are met – the very ones we seek to describe accurately in this enterprise – S knows that *p*. He also argues that this is very likely the best we can do in epistemology. Barry Stroud finds the externalist approach utterly unsatisfying because, in his view, the traditional aim of epistemology is not merely to show that knowledge is possible, but to refute the skeptic's claim that we do not *actually* know anything.[38] It is not enough to say that, if certain conditions are met, S knows that *p*; we want to *show* the skeptic that certain conditions are met, hence that we *do* have knowledge. Stroud claims that, until we do, the skeptic always wins.

Stroud's concerns have a substantial pull on us. On the other hand, so long as we deem the skeptic's basic move, namely, the introduction of a radical skeptical hypothesis, legitimate,[39] it is impossible to prove to the skeptic that we possess knowledge. We would have to prove that our belief-forming processes are conducive to truth, but the skeptic can always say, "But they would not be if you were a BIV, and you do not know that you aren't."

Still, Hill is right that an adequate response to the stronger skeptical argument that knowledge is impossible is no small feat. It is precisely that skeptical argument which is so devastating and has received so much attention since Descartes. A reliabilist response puts the skeptic on the defensive. The skeptic *had* an a priori argument whose conclusion is that knowledge is impossible. If we reply that knowledge is possible if certain conditions are met, the skeptic must show, in order to maintain her thesis, that the conditions are not met – that our belief-forming mechanisms are unreliable – and that is certainly nothing she can do a priori. So while I agree with Stroud that externalism does not give us everything we want – of course it would be *nice* to prove to the skeptic that we know things – I also agree with Sosa that what Stroud wants is simply never forthcoming. (Indeed, Sosa argues that the desire for what he calls a fully general *legitimating* account of knowledge – one where every known proposition is justified by reference to other known propositions, hence one we could use to demonstrate our knowledge to the skeptic – is incoherent on pain of regress or circularity.) Finally, I agree with Hill that undermining the most powerful skeptical argument – against the possibility of knowledge – is a significant and worthwhile achievement.

D. What's missing?

This book concerns specifically modalized epistemology. In Chapter 2 I critically develop process reliabilism. Chapter 3 is devoted to sensitivity and Chapter 4 to safety. Chapter 5 presents the results of all these reflections. There are two very important theory types not discussed in detail here: contextualism and virtue epistemology. One reason is that, while both contextualism and virtue theories often include modal elements, the uniqueness of each hinges on claims that go beyond modalism in ways I believe are unwarranted, and both theories are compatible with various construals of the required modal aspects of belief. Another reason is that neither theory type is essentially externalist; there are internalist versions of both.[40] I shall briefly describe these theory types here and continue the discussion in Chapter 6.

Contextualism (semantic or attributor contextualism) is the view that the truth-conditions of sentences like "S knows that *p*" vary either (1) with features of conversational context; or (2) with the interests or purposes of knowledge attributors.[41] One of the main benefits of contextualism, intimated earlier in this introduction, is that, if true, it achieves anti-skeptical results while not denying the principle that knowledge is closed under known entailment – the principle that, if S knows that *p*, and S knows that *p* entails *q*, then S knows that *q*. The skeptic sometimes employs this quite plausible principle, at least implicitly, to argue that we do not or even cannot know anything. Suppose *q* is *I am not a BIV* and *p* is *I have (real) hands*. If I do not know that *q*, and I know that *p* entails *q*, then it seems I do not know that *p* either. (If I did, it *appears* that I could infer *q* because I know that *p* entails it. Whether the inference can be a means of *acquiring* knowledge that *q* is an important issue that I take up in the first half of Chapter 6.)

Here is a brief example of how contextualism upholds closure compatibly with knowledge. In contexts where skeptical worries are not an issue for the conversational participants considering the knowledge status of some agent, S, it may be true that S knows, for instance, that she is having dinner with friends, and because this entails she is not a BIV and she knows this entailment, S knows she is not a BIV. (Never mind, for now, *how* she knows that. Claiming that one can deduce it from known ordinary propositions seems a reasonable suggestion, but then one would wonder why skepticism ever caused angst. Contextualists pride themselves on taking skepticism seriously, yet I wonder whether even they give it its proper respect.) In contexts where skeptical worries are at issue, even given that S has the same evidence as in the original scenario, it is not true that S knows she is not a BIV – in this context the BIV hypothesis is relevant and S cannot rule it out – but neither is it true that she knows that she is having dinner with friends. It is indeed this combination of features – an anti-skeptical result that conforms to how "knows" locutions are actually used and the preservation of the closure principle – that contextualists claim makes their view superior to the theory I shall develop and defend here. For many contextualists, a criterion of adequacy on any theory of knowledge is that it be able to explain the power of skeptical arguments while not rejecting the intuitively unassailable closure principle, at least not without *explaining* how closure could fail. In the first part of Chapter 6, I take up this challenge, arguing that at the very heart of process reliabilism is the thesis that knowledge demands responsiveness to the world. Because our ordinary (empirical) beliefs are so responsive but the anti-skeptical implications of those beliefs are not, we know the former but not the latter, hence closure fails. I shall also offer reasons for rejecting contextualism.

Virtue theory is an even more diverse terrain than contextualism. What all virtue theories have in common, however, is the basic idea that knowledge derives from the exercise of intellectual virtues, at a minimum stable dispositions of the agent to produce true beliefs. The notion of a *stable* disposition is a modal one, since it implies that the virtue in question must typically produce true beliefs throughout a range of nearby worlds. If it only produced true beliefs in the actual world, perhaps because of some quirky feature, it would not be stable. A strong motivation for virtue epistemology is that, when someone knows, she deserves some sort of credit, whereas one deserves no credit for a lucky true belief. Virtue epistemology thus aims to eliminate epistemic luck and improve upon modal epistemologies, for instance process reliabilism, that make no essential reference to credit due to a knowing agent. (If one's concern is reflective luck, one will think of virtues internalistically, and if one's concern is veritic or process luck, one may think of virtues as nothing more than stable dispositions and not require that the agent understand that or how those dispositions are truth-conducive.) The notion of credit is important, for a well-known argument from Linda Zagzebski concludes that reliabilism, or any truth-conducive characterization of warrant, cannot make sense of how knowledge is more valuable than mere true belief, since on those views attaining true belief is the only recognized epistemic good.[42] But if we think of creditworthiness as an indication of some value accruing to the agent's way of believing, then we can say that knowledge is more valuable than mere true belief because it reflects credit due to the agent.

In the second part of Chapter 6, I respond to the "value problem," arguing that, *whether or not* a reliable process yields truth in any particular instance, we recognize the value of the strong epistemic position of an agent whose beliefs are reliably formed in the requisite way, even in cases where it is dubious to grant credit to the agent for her beliefs. The basic strategy, then, will be to locate a source of value (1) independently of true belief, by showing that a reliable process is valuable even in cases where it fails to achieve the goal of truth; and (2) independently of creditworthiness, by showing that in many cases of knowledge, including many of those where belief is based on stable cognitive dispositions, it is inappropriate to grant credit to the agent. (When the chicken sexer case is described carefully enough to avoid veritic and process luck, at least externalist virtue theorists ought to agree that it constitutes knowledge, yet one might easily decline to give the chicken sexer credit for her beliefs. She is clueless about her belief-forming process.) There is value in the strong epistemic positions that give rise to reliably formed beliefs, even when the target of truth is missed. In short, though this book is somewhat narrowly focused on modal epistemology, I aim, especially in Chapter 6, to treat, albeit indirectly, these two other main trends in current epistemological theorizing: contextualism and virtue theory.

In the meantime, I trust that I have said enough already to indicate what I hope to achieve in this book. It is time now to begin our investigation of modalized epistemology. When finished, the theory I prefer will emerge in Chapter 5, and then we can test its ability, in Chapter 6, to respond to criticisms against it put forth by contextualists and virtue theorists. Let's get started.

2 Process reliabilism

We begin our exploration of modal epistemologies in earnest here with a discussion of global, or process reliabilism. In this chapter I lay out the salient features of Alvin Goldman's reliabilism, which has both global and local features. Our discussion of the evolution of Goldman's thinking on these issues will reveal reasons for including both. Though I shall reserve investigation of the deficiencies of purely local versions of reliabilism[1] for the next two chapters, we will see here (§C) that even the strongest version of global reliabilism faces clear counterexamples that motivate the addition of a local reliabilist element, which in turn sets the stage for the succeeding chapters. As we work through the evolution of Goldman's views, we shall have occasion to remark on their implications both for the Gettier problem, a desired solution of which provides significant motivation in Goldman's early work, and for epistemic luck in general. Toward the end of the chapter I will discuss the implications of Goldman's view for skepticism (§D), and will conclude (§E) with a brief recap of the lessons we should take forward toward an account of a proper modalized epistemology.

Though I make no pretensions to giving an accurate history of the origin and evolution of process reliabilism, it will be useful to have a more or less historically linear narrative of its development. Our hero is Alvin Goldman who, through a series of papers and (for our purposes) *a* book, has created and refined process reliabilism. Following his lead, our point of departure is the Gettier problem as described in the introduction. Recall that Gettier cases can be generated *ad nauseum* by dreaming up examples of double luck. An agent forms a justified belief that happens to be true, but the agent does not know because the ground or justification for her belief has nothing to do with the fact that makes it true. Smith believes on good grounds – for instance, she has been getting rides in a Ford from Jones, and the normally trustworthy Jones said it was her car – that Jones owns a Ford. Smith thus comes to believe that the disjunctive proposition, that either Jones owns a Ford or Brown is in Barcelona, is true. This happens to be true because Brown is in Barcelona, but it is false that Jones owns a Ford. Intuitively, Smith does not know that either Jones owns a Ford or Brown is in Barcelona. It is bad luck that Smith's grounds are false, but good (veritic) luck that her belief in the disjunctive proposition is true.

A. Goldman's causal theory

Essentially, the problem in Gettier cases is that one's grounds for belief are disconnected from what makes the belief true, paving the way for double luck. Goldman's original response to this problem was to forge the requisite connection, in particular a causal link.[2] Take the Gettier case just described. There is no causal link between Smith's true belief – that Jones owns a Ford or Brown is in Barcelona – and the fact that makes it true, namely, that Brown is in Barcelona. Rather, the causal origin of Smith's belief is her perception of the Ford and her being told that Jones owns it. Goldman offers a number of similar cases to illustrate the necessary causal connection.

Perception (HOLOGRAM VASE). Suppose there is an actual vase in front of Jones, but that between it and Jones is a hologram of a vase that blocks her view of the actual vase, and it causes her to believe there is a vase in front of her (Goldman 1967, 359). Her belief is true and, one wants to say, justified because Jones sees what looks like a vase, but Jones does not know because the vase itself is not the cause of her perceptual belief, nor is there (*ex hypothesi*) any other causal connection between the actual vase and Jones's belief.

Inference (LAVA). Suppose a geologist happens upon a heap of lava and, using her background knowledge, infers truly that a nearby mountain has erupted (Goldman 1967, 361). Suppose further, however, that someone removed all the lava from the original eruption, and some years later another person (for whatever reason) deposited a mass of lava with similar characteristics on the same site. Intuitively, the geologist does not know, though she has a justified true belief, that the mountain erupted. Goldman diagnoses this case by saying that the requisite causal link between the geologist's belief and the fact that makes it true is missing. The geologist infers that *this* lava poured forth from *that* mountain, from which she infers that the mountain erupted, but the first inference is incorrect.

Importantly, Goldman does not claim that the fact must be a *direct, antecedent* cause of one's belief, but only that there be an appropriate causal link between the fact and the belief. In this way he can account for, among other things, knowledge of future events (if such there be). If Smith tells Jones on Sunday that she is going downtown on Monday, then, assuming that Smith is trustworthy, Jones may know that Smith is going downtown, but surely the fact that Smith goes downtown on Monday cannot cause Jones's belief on Sunday (Goldman 1967, 365; GOING DOWNTOWN). There is, however, a causal link between the fact and Jones's belief, namely, Smith's sincere assertion that she intends to go downtown the next day; the intention is the cause of the assertion, hence of Jones's belief, and also of Smith's action. The existence of some such "appropriate" causal link is necessary for knowledge, and makes it possible that Smith knows that Jones will go downtown on Monday.

In a similar way, Goldman can account for other cases where one's belief is not a causal consequent of the fact in question. An agent may know that smoke is coming out of her chimney because she sees a fire in her fireplace (Goldman 1967, 366; CHIMNEY). The smoke from the chimney does not cause the belief, but one way may well infer that there is smoke coming out of the chimney from the perceived fire. Here there is an "appropriate" causal connection between the fact that smoke is coming from the chimney and one's belief about that fact – the fire causes both.

Several times already I have used Goldman's term "appropriate." Goldman is well aware that the term is not well defined, but he sees this as a potential virtue of the theory, insofar as the kinds of causal connections that constitute knowledge are currently not fully understood, and thus research into the relevant causal links provides a basis for cooperation between philosophy and the special sciences.

> Leaving the list [of appropriate causal processes] open is desirable, because there may be some presently controversial causal processes that we may later deem "appropriate" and, therefore, knowledge producing. Many people now doubt the legitimacy of claims to extrasensory perception. But if conclusive evidence were to establish the existence of causal processes connecting physical facts with certain persons' beliefs without the help of standard perceptual processes, we might decide to call such beliefs items of knowledge.
>
> (Goldman 1967, 371)

This is one way in which Goldman anticipates Quine's demand for the naturalization of epistemology, that is, for taking the inquiry into the nature of knowledge out of the armchair and into the lab.[3] (Quine would presumably not be all that happy with the present work, whose executor has not been in a science lab since college.)

On the other hand, the extension of "appropriate" causal connections is surely in part a matter for philosophical reflection rather than empirical study. Take again the case of the hologram vase, and now suppose that its existence is due to (caused by) someone perpetrating the following hoax: She wants Jones to form a justified true belief, and so the presence of the actual vase causes her to create the hologram. Now we have a causal connection between the vase and Jones's belief that there is a vase, and we understand intuitively – not on the basis of science – that this connection is *inappropriate*. We can explain the inappropriateness of the causal connection within the framework of Goldman's causal theory. The causal connection is inappropriate because there is no direct perceptual causal connection between the vase and Jones's belief that there is a vase, and Jones does not correctly reconstruct the causal chain from the presence of the vase to her belief.[4] (The causal chain reconstruction problem arose in the lava case. I shall say more about the correct reconstruction requirement for inferential knowledge presently.) The task for science is to explore the possibility of (other) appropriate causal connections, such as, perhaps, extrasensory perception, and to describe the nature of such connections.

Goldman's causal requirement makes it difficult to construct cases where fact and (grounds for) belief come apart as in Gettier examples. Indeed, with respect to inferential belief, Goldman claims that an agent's inferences must correctly reconstruct and hence mirror the causal chain leading from the fact in question to the original beliefs from which she makes inferences. Take Goldman's NEWSPAPER case (Goldman 1967, 363). A newspaper reporter reports that *p*, but the report is misprinted and it says not-*p*. In reading the paper, Jones fails to see the "not" and reads that *p*. She believes truly that *p*, but does not know, because in trusting the newspaper she infers that *p* (though it says not-*p*). Jones's inference does not correctly reconstruct or mirror the causal chain leading up to her reading that *p*. Goldman's additional reconstruction requirement thus eliminates many possible Gettier double-luck cases.

Based on his fundamental insight about the necessity of a causal connection, Goldman offers a detailed analysis of the truth conditions for *S knows that p* (Goldman 1967, 369–70). For our purposes, laboring over Goldman's analysis is unnecessary, since it represents only a precursor to the version of Goldman's process reliabilism in which we will be primarily interested. And though it is worth noting the oft-heard complaint that it is especially difficult to understand how logical, mathematical, and ethical facts could literally be causes of belief, I do not wish to focus on this problem. It seems to me that if Goldman's analysis correctly captured the truth conditions of ordinary empirical knowledge, it would be a substantial achievement.[5]

Alas, though the causal requirement adds something quite strong to the traditional analysis of knowledge while at the same time jettisoning the traditional justification condition, it is not strong enough. Cases examined in Goldman's subsequent discrimination account of perceptual knowledge demonstrate why.

B. Goldman's discrimination requirement and relevant alternatives

Let us begin with the famous BARN FACSIMILE example in Goldman's "Discrimination and Perceptual Knowledge."[6] Suppose Henry is driving in the countryside with his son, points to a barn, and says, "That's a barn." Suppose it is a barn, and that Henry is good at identifying barns; he can at least distinguish barns from silos, houses, sheds, and, for that matter, from barnyard animals. On Goldman's causal analysis, Henry knows that it is a barn, since perception of the actual barn causes this belief.

Suppose, however, that there are many papier-mâché barn facsimiles in the area that look just like real barns, such that Henry cannot distinguish, from the road, real barns from fakes. The barn Henry points to happens to be real, but had it been a fake, Henry would have mistaken it for a real barn. Intuitively, Henry does not know that the object to which he points is a barn because he very easily could have been mistaken. It seems he is lucky to have pointed out a real barn – his belief is veritically lucky.

As the title of the paper implies, Goldman aims to construe perceptual knowledge as a discriminative capacity. He gives up the requirement that a belief must stand in an appropriate causal relation with the fact that makes it true in favor of the claim that, to constitute knowledge, a belief must be formed by a reliable causal process. We thus enter the realm of process reliabilism, where "a cognitive mechanism or process is reliable if it not only produces true beliefs in actual situations, but would produce true beliefs ... in relevant counterfactual situations" (Goldman 1976, 771). With the appeal to counterfactual situations, we hereby also enter the realm of modalized epistemology.

I shall come back to the very large topic of the correct understanding of "relevant counterfactual situations," or relevant alternatives, in just a moment. For now, notice that, in an environment where fake barns are ubiquitous, there is a close counterfactual situation where Henry points out a fake barn and forms the belief that it is a barn. Were that to happen, Henry would form a false belief; that is, he cannot discriminate real and fake barns.

What, then, makes an alternative *relevant*? If the theory is to make room for the possibility of knowledge at all, it cannot count every logical possibility as relevant, since it is precisely our inability to distinguish, for example, dreaming or BIV scenarios from everyday ones that the skeptic relies on to deny that knowledge is possible. Here is Goldman's first comment on this issue: "The presence of facsimiles does not 'create' the possibility that the object Henry sees is a facsimile ... What the presence of the facsimiles does is make this possibility relevant; or it makes us *consider* it relevant" (Goldman 1976, 775). Over the years, Goldman has considered a number of related candidates for the correct account of what makes an alternative relevant. Note, however, that there are (at least) three general classes of relevance determinants suggested by the last quotation: (1) relevance is determined solely by environmental factors independently of what anyone deems relevant; (2) relevance is to be decided by what *we as theorists of knowledge consider relevant*, based on the objective features of the environment (for example, on the presence of fake barns in the area); (3) relevance is a matter of what *knowledge attributors consider relevant*, presumably depending on their attitudes, intentions, purposes, or the conversational context. The third class includes the varieties of the recently much discussed view called (semantic or attributor) contextualism.[7]

It is likely that by "us" Goldman means to refer to theorists rather than to participants in each particular scenario at issue, in which case it would be incorrect to attribute to him an anticipation of contextualism, a view he does not accept. But here I am merely exploring logical possibilities. (Of course, the classes described under (1) and (2) also have a contextual element, but not one that hinges on features of conversational context or the intentions and purposes of the attributor. I shall reserve the term "contextualism" for the third class only. As I said in the introduction, I shall discuss contextualism only indirectly, especially in Chapter 6, when it raises specific issues and problems for other theories.) In addition, it is difficult to locate a substantive difference between (1) and (2), because at the end of the day, as theorists of knowledge we seek an account of relevance, from which it follows that even if (1) were the right approach, we still want to understand how environmental factors determine relevance, and any such explanation will necessarily involve considering some possibilities relevant to the exclusion of others. Setting contextualism aside, then, what are the factors that determine which counter-possibilities one must be able to rule out if one's true belief constitutes knowledge?

This is an important question both for an understanding of the modal features of process reliabilism and for issues that arise in the following chapter. Thus though I shall not attempt to answer the question fully here, it will be useful to introduce the pertinent issues and to get as precise as possible Goldman's view of relevant alternatives. Fred Dretske uses the notion of relevant alternatives to argue that knowledge is not closed under known entailment.[8] He argues that knowledge of some propositions involves a presupposition that other propositions are false. Here's a brief version of his famous CLEVERLY DISGUISED MULE example. Suppose at a zoo you are at the zebra pen, which is clearly marked "zebras," where you see animals that look like zebras, and you can tell the difference between zebras and other animals typically found in a zoo. Do you know that the zebra is not a cleverly disguised mule? On the assumption that you cannot tell a real zebra from a clever fake, it would seem that you do not. On the other hand, such a possibility seems farfetched, or irrelevant to whether you know that the animal is a zebra. Here's the situation:

(i) You know (or could know) that it is a zebra.
(ii) You know that, if it is a zebra, it is not a cleverly disguised mule.
(iii) You do not know that it is not a cleverly disguised mule.

The compatibility of these three propositions – which just *is* the denial of closure – is admittedly counterintuitive, and for some constitutes a very strong reason for rejecting an account like Dretske's. Given the controversial nature and complexity of the issues surrounding closure, I shall set aside treatment of the issue until later, in Chapter 6.

Dretske argues that epistemic operators such as "knows that" do not always penetrate (or go through) to presuppositions that are entailed by what one knows, where in the cases that interest us what is presupposed is the denial of some irrelevant alternative, for example, that the animal is a disguised mule. He then owes us an account of presupposition, and much that he says in "Epistemic Operators," for instance in his discussion of the locution "explains why" [S did action *a*], suggests that presupposition is a function of the context. Suppose we want an explanation for why Brenda did not order dessert (Dretske 1970, 1020; BRENDA). That she did not order dessert *entails* that she did not order it and throw it at the waiter, but the explanation for her not ordering dessert, perhaps that she is on a diet, does not also *explain why* she did not order dessert and throw it at the waiter. (An explanation of that might be that Brenda is afraid of looking like a fool.) What counts as an explanation depends on what one wants explained, which is a feature of the interests of the parties involved. The point is, then, that explanations do not penetrate to logical entailments of what they explain, precisely because those entailments are sometimes irrelevant. Dretske argues that knowledge, in like fashion, does not always penetrate to known entailments, and for the same reason.

The problem with this approach to relevant alternatives, however, is that it can explain closure failure only if we imagine a change in context when evaluating the truth or falsity of (i) and (iii). That is, Dretske can say that (i) is true because the possibility that the animal is a cleverly disguised mule is irrelevant in ordinary contexts, but (iii) is true because in evaluating it the clever fake possibility *is* relevant. Stine argues persuasively that the implied change of context constitutes an equivocation – there is no *one* context in which both (i) and (iii) are true.[9] The truth of (iii) implies that the clever fake possibility is relevant, but in such contexts (i) is false. (Analogy: It is not surprising that the conclusion of the following argument does not follow from the premises. Today is Tuesday [spoken on a Tuesday]. I always go to the dollar movie theatre on Tuesdays. Therefore, I will go to the dollar movie theatre today [spoken on a Wednesday]. Surely Dretske intended his denial of closure to be more interesting than this.)

There are at least two distinct revisions to (this interpretation of) the original Dretskean line that avoid the equivocation charge. The first is to accept contextualism and require that the context be held constant in assessing relevant alternatives. One nice implication of this strategy is that it makes room for the possibility of knowledge without denying closure. In contexts where skeptical possibilities are relevant, one does not know that ordinary propositions are true and one does not know the denials of skeptical hypotheses; where skeptical possibilities are not relevant, one can know ordinary propositions *and* the denials of skeptical hypotheses.[10]

Neither Dretske nor Goldman ultimately follows the contextualist line of thought.[11] Though Goldman himself is not committed to any specific view of relevance, he seems to agree with the implication of Dretske's revised view that relevance should be construed in modal terms and without appeal to what is deemed relevant by conversational participants or knowledge attributors.[12] In particular, Dretske argues that the proposition *S knows that* P *on the basis of reason* R entails that "R would not be the case unless P were the case" (Dretske 1971, 1). Dretske makes clear that relevance is determined by objective features of the environment, which is consistent with Goldman's suggestion that it may simply be the *presence* of facsimiles that makes the possibility that Henry is looking at a fake relevant. If there were no fakes within hundreds of miles of Henry, whether Henry knows it or not, the possibility that the thing Henry sees is a fake, hence the worry that he would not be able to tell the difference, might well be irrelevant. In such an environment, if the thing Henry points to were not a barn, he would not believe it is because there is nothing in the area that he would mistake for a barn. (Put in the terms of Goldman's discrimination account of perceptual knowledge, there are no relevant perceptual equivalents of a barn that Henry would mistake for the real thing.) Intuitively, in that environment, if the object in question were not a barn, it would not have been a barn facsimile, and so the (theoretical) worry that it is may be deemed irrelevant.

Dretske's main claim about conclusive reasons (R), ones that yield knowledge, can be put in the form of a conditional, thus: If P were not true, R would not be the case. For example, if the barn pointed to were not real, Henry would not have his reason for thinking it is. (The point is that, as the case is described, Henry *would have* this reason, and thus his reason is not conclusive, which explains why he does not know.) This is very close to Nozick's subjunctive conditional theory, which states that the following condition is necessary for knowing that *p*: If *p* were not true, S would not believe that *p*. On the assumption that Dretske's and Nozick's necessary conditions are not relevantly different for present purposes, the obvious demand here is a story about how to determine what would be, or would have been the case, in counterfactual circumstances. Dretske insists that, when assessing the truth-value of the conditional, we must hold fixed all the circumstances in the particular, actual situation that are logically and causally independent of *p* (Dretske 1971, 7–8). In the fake barns case, the fact *that there are fake barns* is logically and causally independent of the proposition that Henry believes: That [demonstrated object] is a barn. Holding that fact fixed, it turns out that if Henry's belief were false, he would still have his reason for thinking it true, because he would be looking at a barn facsimile. Nozick's theory gets the same result – if the proposition Henry believes were false, he would still believe it.

Given that Goldman, Dretske, and Nozick agree that relevance is determined by objective features of the environment, the remaining question is the extent of the space of possible worlds that are relevant, or how many counter-possibilities are relevant. In his book, *Epistemology and Cognition*, Goldman rejects Nozick's "pure" subjunctive theory (and, by implication, Dretske's conclusive reasons theory), which he distinguishes from his own relevant alternatives account, insofar as the latter allows more counterfactual situations to be potentially relevant to the question whether one has knowledge. Though I shall treat Nozick's theory, and to some extent Dretske's, in more detail in the following chapter, it will be useful to get a grip on Goldman's rejection of the pure subjunctive view of relevant alternatives here.

Suppose a parent takes a child's temperature using an accurate thermometer and, reading the thermometer, forms the belief that the child's temperature is 98.6 degrees (Goldman 1986, 45; THERMOMETER). If the proposition believed were false, the parent would not believe it. But suppose there are many thermometers in the medicine chest, and the parent happens to select the only accurate one. Would we say that the parent knows the temperature? Goldman says "no," and he thinks that this is a problem for the pure subjunctive view because it does not take into account enough counterfactual possibilities. Goldman's criticism implies that the pure subjunctive view is committed to saying that the parent does know, which suggests the following interpretation of Nozick's subjunctive conditional: "If *p* were false, S would not believe that *p*" is true if S does not believe that *p* in *the nearest possible world* where it is false. In the *nearest* possible world where it is false that the child's temperature is 98.6 degrees, the parent uses the same (accurate) thermometer as that used in the actual world. Goldman's objection to the pure subjunctive view is that there are *other* relevant alternatives, for instance, the counter-possibility where the parent selects an inaccurate thermometer. If she had, she would have formed a false belief, and that is why she does not know. (Goldman recognizes that his account of relevance is not very precise. He does, however, say about relevant counterfactual situations that they are "ones that are highly 'realistic', or conform closely to the circumstances of the actual world."[13])

Generally speaking, the recent history of attempted analyses in epistemology shows that appeal to intuition in deciding whether a particular case constitutes knowledge is a reasonable strategy, at least to the extent that most cases garner widespread agreement. But, of course, there are exceptions, and this case may very well be one of them. The parent uses an accurate thermometer and forms a true belief. It is at least not obvious that she does not know. The example *appears* to be formally analogous to the barn facsimile case: It was merely lucky that Henry pointed to an actual barn, and "it was just luck that a good thermometer was selected" (Goldman 1986, 45). There is one difference, however, and whether it is telling I leave to the reader to decide. Henry cannot distinguish, by eyesight from the road, a real barn from a fake barn, whereas the thermometer used *does* distinguish temperatures, and this difference is born out by the subjunctive account. Were it false that the object seen is a barn, Henry would still believe that it is a barn, but were it false that the temperature is 98.6 degrees, the parent would not believe that it is 98.6 (on the present, "pure" reading of subjunctives), as Goldman notes.

Now, surely it is question-begging to appeal to the truth of the latter subjunctive conditional to support the pure subjunctive view, since Goldman's objection is precisely that the pure reading of the subjunctive makes it *true* and yet the parent does *not* know, but that is not exactly what I am doing. Rather, my point is that the thermometer case is in fact different from the barn case, and this opens the possibility that Goldman's intuition about the thermometer example is not compulsory. Perhaps we should distinguish levels of knowledge, setting the minimum threshold at being able to rule out alternatives "defined" by the pure subjunctive conditional.[14] (As we shall see, that some subjunctive conditional holds is by itself not sufficient for knowledge, and some argue that it is not necessary. We'll come to those issues in the next chapter, but here let us focus on which alternatives one must be able to rule out in order to know.) When one is able to rule out even more alternatives, one knows even better. For now, I shall leave this as the merest of suggestions, though one I find promising. (I want to add, however, that my intuitions here seem to accord with Dretske's. See his 1971, 8–9.)

If that proposal seems implausible, there is another reply one could make on behalf of the subjunctive conditional view. It is that the parent's belief indeed *fails* to satisfy the conditional because the world where she selects a faulty thermometer is otherwise exactly like the actual world. Given that she very easily could have used a bad thermometer, it is not obviously false to say that, were the temperature not 98.6, she would have believed it anyway. Intuitively, in the *very* closest world where the temperature is not 98.6, she would not believe that it is, because the closest world is one where *only* the temperature is different. But there are a number of *very close* worlds where the parent uses a faulty thermometer and forms a false belief. The present suggestion may very well be inconsistent with Dretske's idea that everything logically or causally independent of p – the temperature is 98.6 – must be held constant in assessing the counterfactual; were p false, the parent would not believe it, because the thermometer actually used must be held constant. And perhaps this opens a logical space between Nozick's and Dretske's interpretations of the subjunctive conditional (where only the latter would be "pure"). In any case, the point is that a subjunctive account of relevant alternatives may accord with the intuition that the parent does not know the temperature, because there is a plausible reading of the conditional that says, if the child's temperature were not 98.6 degrees, the parent would have believed it is anyway (assuming, say, that the inaccurate thermometers are "stuck" on 98.6).[15] As I indicated in the introduction, there are cases, and this is one, where intuitions about what would be true in some counterfactual circumstance diverge. Clarifying, though not settling, this conflict is achieved by noting that different interpretations of modal conditionals hinge on different views of which possible worlds are relevant to assessing those conditionals.

Hopefully this discussion has illuminated both the central features of Goldman's understanding of relevant alternatives and some of the main options available to the relevant alternatives theorist. Goldman is not a contextualist, and he does not accept the view that only the very closest counter-possibility is relevant. I have added that, (1) even if one were to accept the pure subjunctive conditional view, as Goldman understands it, putative counterexamples might be handled by claiming that, so long as one's belief does not violate the conditional, it constitutes at least minimal knowledge, and I have suggested that (2) the subjunctive conditional view may in fact be compatible with a consideration not only of the nearest possible world where not-*p* but of other close worlds that are almost exactly like it. (If this is right, then the correct reading of subjunctives requires rejecting the pure interpretation.) I shall defer further discussion of this topic until the next chapter, where Nozick's view will come into focus. Let us return to our discussion of Goldman's discrimination account.

Technicalities aside, then, Goldman's analysis of perceptual knowledge is that S perceptually knows, of an object *a*, that it has property *F* if and only if:

(i) *a* has *F*;
(ii) *a*'s having *F* causes S to have the percept that *a* has *F*;
(iii) S's percept that *a* has *F* causes S to believe that *a* has *F*; and
(iv) there is no perceptually indistinguishable relevant alternative that would cause S to form the false belief that *a* is *F*. (For the precise statement, see Goldman 1976, 785–6.)

Goldman offers the following summary of his analysis:

> What our analysis says is that S has perceptual knowledge if and only if not only does his perceptual mechanism produce true belief, but there are no relevant counterfactual situations in which the same belief would be produced via an equivalent percept and in which the belief would be false.
>
> (Goldman 1976, 786)

Let us pause to note some salient features of Goldman's account. First, it is not obvious that this is a version of global, or process reliabilism, insofar as it appears to focus on the modal status of particular beliefs, and thus seems to highlight a local reliability requirement. (The impression is amplified by the foregoing discussion of relevant alternatives.) This may suggest that we're working under the wrong heading, *process reliabilism*, in this chapter. However, it is clear from the passage quoted earlier, taken from the beginning of his paper, that Goldman aims to emphasize the reliability of processes generally, which is a function of their total belief outputs: "a cognitive mechanism or process is reliable if it not only produces true beliefs in actual situations, but would produce true beliefs ... in relevant counterfactual situations" (Goldman 1976, 771). In addition, as we shall see in our discussion of his refined account, Goldman wants to emphasize the global reliability of belief-forming processes. This is in part because he believes that local reliabilism by itself is insufficient. I shall reserve discussion of the problems afflicting local reliabilism for the next two chapters. On the other hand, Goldman claims that global reliabilism is, by itself, also deficient. An example from Linda Zagzebski will clearly demonstrate that global reliabilism is insufficient (see below), and its upshot will be instructive for our ultimate aim of articulating a satisfactory modal epistemology. In the meantime, we need to make explicit the process reliabilism nascent in Goldman's causal and discrimination theories.

C. Process reliabilism and why it is not enough

In this section we are no longer considering counterfactuals that apply to particular beliefs, so local reliabilism is not the relevant issue. That frees us to consider varieties of global reliabilism. En route to Goldman's considered account, let us start with a stripped-down version, namely, actual world process reliabilism.[16] According to this view, knowledge is true belief formed by a process that is reliable (truth-conducive) in the actual world.

A famous counterexample to actual world reliabilism is Plantinga's BRAIN LESION case, where an agent has a brain lesion that causes her to believe that she has a brain lesion, though she has no other evidence for it (and perhaps even some evidence against it).[17] Assuming that such a brain lesion is rare and something of a fluke, hence would not cause true brain lesion beliefs in nearby worlds, it seems rather implausible to say that the agent knows she has a brain lesion. If one is hesitant to call this an example of a genuine process (perhaps because, though the lesion may produce many distinct belief tokens, they are all of one type), there is the BENEVOLENT DEMON.[18] Suppose there is a demon who arranges reality such that, whenever I form a belief based on wishful thinking, which, let us stipulate, I am prone to do, the belief turns out true. Suppose the actual world is like that; my friendly demon will, as soon as I form them, make all my wish-based beliefs turn out true (though I have no clue that the demon exists). If we assume, as I am, that the demon is helping me on a whim (say, for example, because it likes black cats and I happen to have one), such that in most nearby worlds (where I don't have a black cat, and many others where I do) it does not bother to help at all, then it follows that basing belief on wishful thinking is truth-conducive only in the actual world (or, at most, in the actual world and in only some small set of nearby worlds). My beliefs are all too luckily true. This is an example of *process* luck, because the process by which my beliefs are formed produces a high ratio of false beliefs in most nearby worlds.[19] Such examples are admittedly fanciful, but they are not so convoluted as to inhibit verdicts as to whether an agent knows, and we can at least imagine them. They show that actual world reliability, or propensity to produce truth, of a belief-forming process, together with true belief, is not sufficient for knowledge.

This sort of problem prompts Goldman to refine his preliminary description of process reliabilist justification, which states that the "justificational status of a belief is a function of the reliability of the process or processes that cause it" (Goldman 1979, 10). Before attending to Goldman's refinements, a note on his project in this paper is in order. His overall aim is to provide an analysis of some version of justification that is traditionally thought to be a necessary condition for knowledge, but is not committed to Cartesian internalist justification. Thus he is not attempting an analysis of knowledge per se, but only of one of its alleged components. I think a potential danger lurks here, one that I mentioned at the beginning of the introduction. So long as one feels compelled to analyze justification, there will be cases that motivate unduly restrictive necessary conditions on knowledge, some of them threatening a return to internalism. Perhaps this is simply because the word "justification" is so loaded, and there is no escaping its entanglements once one attempts to snare it. I am certainly not denying that a theory of justification is an important epistemological undertaking, but I am suggesting that the better one's account of that notion, pregnant as it is, the likelier it is that justification will not be a necessary condition for knowledge.

A first manifestation of the justificationist snare appears in Goldman's response to the problems of actual world reliabilism. In *Epistemology and Cognition*, Goldman aims to define justification in terms of systems of justification rules. The question relevant for our purposes is how to amend the view that the rightness of the rule system is determined by truth-conduciveness in the actual world, because we have already seen in the brain lesion and benevolent demon cases why this won't work. I shall quote his proposed solution at length:

> The answer that best accords with our intuitions ... runs as follows. We have a large set of common beliefs about the actual world: general beliefs about the sorts of objects, events, and changes that occur in it. We have beliefs about the kinds of things that, realistically, do and can happen. Our beliefs on this score generate what I shall call the set of *normal worlds*. These are worlds consistent with our *general* beliefs about the actual world. ... Our concept of justification is constructed against the backdrop of such a set of normal worlds. My proposal is that, according to our ordinary conception of justifiedness, a rule system is right in any world W just in case it has a sufficiently high truth ratio *in normal worlds*.
>
> (Goldman 1986, 107)

This certainly solves the problems of the benevolent demon and the brain lesion. We do not take such worlds to be normal – to conform to our general beliefs about the actual world. That is why, in such cases, the agent is not justified and hence does not know.

I do not doubt that, as a theory of justification, this is on the right track. After all, even if I am a BIV or my beliefs are uniformly false because I am deceived by an evil demon, it seems plausible to say that I can have justified beliefs, for example, those based on clear and distinct perceptions and cogent reasoning. (This is the NEW EVIL DEMON problem for reliabilism.[20] On the other hand, one might argue that no such beliefs are justified, if one thinks of justification as essentially a link between belief and truth. See §C.2 of the introduction.) But therein also lies the problem – I can have justified beliefs if they are truth-conducive in normal worlds but not at all truth-conducive in the actual world if the actual world is unlike normal worlds. Now suppose that I am justified in just this circumstance; the world is very unlike how I take it to be. Suppose further, however, that the evil demon happens to make a couple of my beliefs true. Thus I have a few justified true beliefs, but it is utterly implausible to say that I know these things. Like the other cases, this extension of the new evil demon problem is fanciful, but it highlights the fact that we ought not to abandon one of the truly ground-breaking insights, *Goldman's own*, of reliabilism, especially as a reply to the Gettier problem and, indeed, to skepticism – that knowledge is constituted by some real connection between belief and truth. Goldman's attempt to capture the intuitive notion of justification leads him away from this point, potentially severing the connection between truth and justification. (This is also the basis for the worry, expressed above, that attempts to capture adequately an intuitive notion of justification may threaten a return to internalism. Goldman himself expresses this possibility.[21]) I want to say, so much the worse for the notion of justification. I will return to what I believe is a better reply to actual world reliabilism (than the normal worlds approach), but first I want to digress to describe another manifestation of the justificationist trap.[22]

Goldman (1979, 18–19) asks us to suppose that Jones is told by his parents, whom he has good reason to trust, that all of his childhood memories are false, fabricated (TRUSTWORTHY PARENTS). He continues to believe that his memories are true, and they are, even though he has no reason not to trust his parents. Intuitively, Goldman says, Jones's beliefs are not justified. Thus Goldman adds to his account of justification a condition that there is no *other* reliable process which, had it been used, would have resulted in one not believing what one does based on the reliable process actually used (Goldman 1979, 20).[23] Jones does not satisfy this condition because, except for claims about his past, his parents are trustworthy, hence, one might say, reliable sources of belief.[24] Violating Goldman's condition, Jones's belief is not justified. (This example raises the generality problem for reliabilism, discussed below and treated in Chapter 5. What is the belief-forming process that Jones might or could have used? Is it basing belief on whatever his parents say? Or is it basing belief on what his parents say when they are not aiming to deceive? Is the latter even a process at all? This sort of case also arises in Chapter 3, §C.2.)

Perhaps this is the right emendation for an account of justification, but then it is not clear that justification, thusly construed, is required for knowledge.[25] It seems to me that one could persuasively claim that Jones does know – he has true memory beliefs, and his memory is reliable. There is always *some* reliable process that might have been used and which could undermine some of one's actual beliefs (there is always some smart, very persuasive person who may *wrongly* disagree with me on a particular point), and perhaps this does create havoc with justification, but if we then require justification, thus suitably refined, for knowledge, many of the hard-fought gains of reliabilism will be lost. Again, this is to fall into the justificationist trap. Reliabilists should get out of the business of justification and stick to the central issue: knowledge. (This is a point I made briefly in the beginning of the introduction and in the discussion of externalism. Another way to put this point is to say that reliabilists should be concerned with veritic and process luck but not with reflective luck.) Of course, some belief-truth link is essential to an account of knowledge, which is exactly what the reliabilist aims to describe. To that end, we need to change the name of the game from justification to the more general concept of warrant.

In order to account for the new evil demon problem within a generally reliabilist framework, while also avoiding some of the pitfalls of the normal worlds approach, Goldman later distinguishes two kinds of justification: strong and weak (Goldman 1988). Roughly, a belief is strongly justified if formed by a truth-conducive process or method, and (merely) weakly justified if, though not strongly justified, the agent is blameless in forming the belief in that she does not believe the process or method used to be unreliable and has no available way of determining that it is.[26] Thus a believer, in the new evil demon case, can be said to be weakly but not strongly justified. Again, this may very well be a plausible refinement of the notion of justification, but it is dubious that weak justification is a necessary condition for knowledge.

The serious pitfalls of the normal worlds approach and appeals to weak justification are brought to light in cases where a belief-forming process may be extremely reliable in some world that is very unlike normal worlds (that is, those worlds very like the one we generally believe the actual world to be). Suppose that clairvoyance is reliable, but is widely believed not to be (Goldman 1988, 62). Imagine that the actual world is like this. Agents who form true beliefs through clairvoyance would seem to be capable of attaining knowledge, and the notion of strong (reliabilist) justification captures this, so long as it is not defined by reference to normal worlds, regardless of whether the agent is weakly justified.

I am not claiming that, if clairvoyance-based beliefs turn out true in *any* world, then clairvoyance constitutes a basis for warrant in that world. We can adapt the benevolent demon case to see why. Suppose I have visions of future events and form beliefs that they will occur on that basis. The demon monitors my visions and actualizes those events, making my beliefs true. As before, if my demon helps only in some worlds (say, where I have one of those black cats he likes), then surely my clairvoyance-based beliefs are not truly reliably formed. Consideration of this kind of case leads us at last to the final amendment to global reliabilism, one that Goldman barely mentions and is hesitant to adopt, but one that I think we should endorse. It is that a belief-forming process is reliable only if truth-conducive not only in the world in which it is employed, for us, the actual world, but also throughout a range of nearby possible worlds. Let us call this modalized process reliabilism.[27] (Goldman: It "seems advisable to assess [a rule system's] rightness in *W* not simply by its performance in *W*, but by its performance in a set of worlds very close to *W*" (Goldman 1988, 63). I agree.) The general idea is also accepted in recent versions of agent reliabilism and virtue epistemology, both of which hold that a reliable process is one grounded in stable cognitive (or intellectual) dispositions of the agent.[28] It does a good job of eliminating process luck, that is, of precluding cases where a belief is true but formed by a process that produces many false beliefs in nearby worlds. Surely the phrase "set of worlds very close to *W*" could use refinement, but the present account of global reliability suffices for now, because no matter how we sharpen the phrase, the following example shows that reliably formed true belief alone does not entail knowledge.

The example is Zagzebski's case of MARY and her husband.[29] Mary has good eyesight and can recognize her husband in the living room, even in somewhat dim lighting. It so happens that her husband's brother, who looks a lot like her husband, is sitting in the husband's usual chair, and Mary forms the belief that her husband is in the living room. Suppose further that her husband *is* in the living room, sitting where Mary does not see him. Her belief is true and formed by a reliable process, one that is reliable not only in the actual world, but throughout the range of nearby possible worlds.[30,31]

The example shows, without a doubt, that global reliabilism is either false or must be supplemented by some other conditions.[32] In the next chapter, we will focus on Nozick's sensitivity condition, the subjunctive conditional requirement described above (§B) in the context of the debate about relevant alternatives. This may handle the Mary case, since if her husband were not in the room, she would still believe that he is, and that is why she does not know. But we will see there that sensitivity has problems of its own, in particular with process luck. In Chapter 4, we'll look at the safety principle (endorsed by Sosa, Pritchard, and others), which holds that a true belief is knowledge only if, in all (or most or many, depending on the version) nearby worlds where the belief is held, it is true. Like the sensitivity condition, safety is (or seems to be) violated in the Mary case, and perhaps this explains why she does not know that her husband is in the living room. Alas, there are problems with safety, too, and thus my preferred version of modalized epistemology emerges in Chapter 5 as the result of this dialectical process.

We have seen how modalized process reliabilism handles many but not all cases of epistemic luck; the Mary problem, as we'll henceforth call it, is a case of veritic luck that reveals the deficiencies of global reliabilism. There are two other problems stemming from process reliabilism that will demand our attention in Chapter 5, which I should like to describe briefly here.

Goldman (1986, 93ff.) distinguishes belief-forming processes and methods, where a process is part of our innate cognitive endowment and a method is a learned or learnable procedure for forming beliefs. (Though the first problem attaches primarily to methods, I will sometimes use the term "process" to cover both processes and methods, in Goldman's sense, because we will later have occasion to distinguish these *general* belief-forming mechanisms from the *specific* methods that attach to particular beliefs.)

Suppose a true believer forms her belief by a process that is in fact reliable – truth-conducive in the actual world and throughout nearby worlds – but uses the process only haphazardly. For instance, she uses a correct long division algorithm, but just happens upon it. As in the Mary problem, her true belief seems just *too* lucky to count as knowledge. Agent reliabilism and virtue accounts like Sosa's can account for lack of knowledge here, insofar as the process used is not part of a stable character trait of the agent's, which constitutes a reason for moving away from process reliabilism to agent reliabilism. Surely this problem puts a constraint on a correct process reliabilist theory, but I shall argue that the requirement that belief be formed from a stable character trait to constitute knowledge is too restrictive. That topic will have to wait until Chapter 5. For ease of reference, we can call this the SERENDIPITOUS RELIABLE PROCESS problem (not to be confused with the serendipitous brain lesion, as it is sometimes called, since the brain lesion is not truth-conducive in nearby worlds).

The second problem is the well-known generality problem for reliabilism.[33] How narrowly ought we to define the process types in question? If they are typed too broadly, such that vision itself is considered a type, then for anyone who happens to believe everything she sees, at whatever distance and in any conditions, the process will be deemed unreliable, and if process reliability is a necessary condition for knowledge, it follows that she does not know anything based on vision. (Perhaps that is not entirely implausible for someone so willing to believe, but it seems a bit drastic.) On the other hand, if they are typed too narrowly, say, for instance, vision of the (demonstrated) cat on the (demonstrated) mat at 1:15 p.m. MST, Friday, January 5, 2007, then, if the belief is true, the process is batting a thousand, and is deemed reliable. But this process is not at all *general*. Clearly both of these types are wrongheaded, but without some answer to the generality problem, the reliabilist does not really have a theory, since she can adopt whichever type suits her needs, ad hoc. But surely also there is some commonsense middle ground here, even if imprecise. Goldman suggests that, to avoid the narrowness problem, the process must be content-neutral, which effectively eliminates ad hoc excess specificity (1979, 12). I will argue in Chapter 5 that, so long as we accept content-neutrality, we are free to individuate processes very narrowly and still avoid the ad hocness complaint.[34] I will leave refinement of these somewhat breezy suggestions until then. There I will argue anew that the requisite generality of processes necessitates an additional local reliabilist condition on knowledge to eliminate veritic luck.

D. Implications for skepticism

In §B we looked at the problem of defining relevant alternatives. Notice, though, that regardless of how and where we ultimately draw the line between relevant and irrelevant alternatives, it is plausible to say that, in the barn case as originally described, Henry does not know because in that environment fake barns abound. This makes it extremely lucky that Henry points to an actual barn, and surely knowledge is incompatible with that much luck, which suggests that the fact that he easily could have pointed to a fake is a relevant alternative.

In addition, if Goldman's analysis is to make room for the possibility of perceptual knowledge, then again, regardless of how and where we determine relevance, some possibilities, for instance radical skeptical ones, must be deemed *irrelevant*. If we think of knowledge as true belief where the believer can rule out all the relevant alternatives, then the possibility of knowledge requires that the skeptical scenarios one cannot rule out are irrelevant.

Let's put this into the form of process reliabilism as tentatively accepted above, subject to later refinements, where knowledge is true belief formed by a reliable process, and where reliability is truth-conduciveness in the actual world and nearby possible worlds. Skeptical scenarios are irrelevant to whether I know that ordinary propositions are true because, presumably, they appeal to very distant worlds – worlds very unlike the actual world – such as where I am constantly deceived by an evil demon, or I am a BIV. Now, of course, since I cannot rule out those possibilities, on this view I do not in fact know that the actual world is not a BIV world. Hence process reliabilism does not establish that knowledge is actual, but does establish that knowledge is possible, and I for one think that is a worthwhile achievement. After all, as we saw in the introduction (§C.3), one of the most powerful skeptical arguments aims to show that knowledge is impossible because we have no evidence that rules out the skeptical hypotheses, and if we do not know that such hypotheses do not obtain, it appears to undermine the claim that our ordinary beliefs are justified.[35] Process reliabilism answers this kind of skepticism by appealing to the (plausible) possibility that some of our belief-forming processes are reliable – that they allow us to distinguish truths about the actual world from many counter-possibilities. The skeptic has no a priori argument that our belief-forming processes are not reliable. To that extent process reliabilism constitutes significant progress in the theory of knowledge.

E. Conclusion

This chapter sketched the history of process reliabilism by focusing on the seminal work of Alvin Goldman. We have arrived at a version of the theory that eliminates many of the most egregious cases of epistemic luck, especially process luck, without being lured into the trap of justificationism. But we have also seen, in the Mary problem, that global reliabilism is not enough because it does not preclude veritic luck, and the appeal to stable cognitive characteristics is impotent to help us with this. This motivates an exploration of local versions of reliabilism, that is, of modal principles aimed at particular beliefs, together with an honest inquiry into the sufficiency of those accounts. The theory I shall ultimately defend will be rather like Goldman's but with (1) an explicit defense of a subjunctive conditional (sensitivity) component; (2) an explicit endorsement of actual-plus-close worlds account of reliability; and (3) a denial of the claim that justification, in any sense that bears close relation to the intuitive notion (as exemplified in normal worlds reliability, weak justification, or internalism broadly construed), is required for knowledge. With respect to the first two aspects of my theory, I will argue that sensitivity is not merely a necessary supplement to process reliabilism in order to avoid the Mary problem, but that sensitivity and process reliabilism are naturally complementary, and, taken together, they can provide an independently motivated solution to the generality problem for reliabilism and an explanation of the otherwise counterintuitive failure of the closure principle.

We have also seen that any version of process reliabilism faces other serious challenges, including the Mary problem and the serendipitous reliable process problem. All of these issues will be taken up after our discussions of local reliabilism, which constitute the next two chapters.

3 Sensitivity

As we learned from the Mary problem in Chapter 2, the deficiency in process reliabilism, even when modalized so that reliability entails truth-conduciveness in the actual world and in nearby possible worlds, is that a true belief can be formed by a reliable process and at the same time not be responsive to the particular fact that makes it true.[1] Mary has a true belief, formed by her modally reliable vision but based on her mistaking her husband's brother for her husband, that her husband is in the room, but she does not know that he is in the room. The clearest way of expressing her lack of knowledge is to say that, if her husband were not in the room, she would believe that he is. The obvious next step is to incorporate the lesson drawn from the Mary problem into our theory. Nozick's theory of knowledge does at least this much by adding a sensitivity condition to the analysis of knowledge. In this chapter, we will set out Nozick's theory and focus almost exclusively on the sensitivity condition. The most powerful objections to sensitivity claim that it is too strong, wrongly yielding a verdict of *no knowledge* in many cases.

Nozick's epistemology makes little mention of the process types through which one comes to form a belief. Appeal to the method by which the specific belief is formed does play a crucial role for Nozick in his response to potential criticisms, which I shall explain (§B). But the present point is that Nozick's modalized epistemology is quite distinct from process reliabilism, and so can and should be treated separately. Whereas modalized process reliabilism is concerned with the modal status of belief-forming processes, Nozick's theory is concerned with the modal status of particular beliefs, and thus is a version of local reliabilism.

Nozick, like process reliabilist theorists, is keen to solve the Gettier problem and to eliminate (especially veritic) epistemic luck. And like Goldman's original causal account, the aim is to ascertain the requisite link between a belief and the fact that makes it true that is constitutive of knowledge. In the Gettier cases, as we have seen, one's true belief is justified but only luckily true. It could easily have been false even with the believer's grounds for belief intact. Nozick's strategy is to replace the traditional justification condition with conditions ensuring that, in cases of knowledge, one's true belief cannot float free of the relevant fact – the belief must somehow track the truth. After setting out Nozick's basic view and running it through some test cases (§A), we will make an important refinement to it (§B). This will take us to a more challenging assessment of its adequacy (§C), where we will encounter objections, most notably the aforementioned complaints that the theory is too restrictive, discounting genuine cases of knowledge. In §D, I will summarize the general conclusions from this chapter.

A. Nozick's subjunctive conditional theory of knowledge

Nozick's first two conditions for knowledge are the standard ones. S knows that *p* only if:

(1) *p* is true.
(2) S believes that *p*.

Nozick's task is to provide the further necessary conditions which, taken together with (1) and (2), suffice for knowledge. The other conditions take the form of subjunctive conditionals, which Nozick abbreviates thus:

(3) not-*p* $\rightarrow$ not-(S believes that *p*).
(4) *p* $\rightarrow$ S believes that *p*.

(3) is to be read as, if *p* were not true, S would not believe that *p*. (Consistently with other authors' usage, I am calling this the sensitivity condition, though Nozick thinks of satisfaction of (3) and (4) together as constituting sensitivity; he says that "Condition 3 is a variance condition, condition 4 is an adherence condition.")[2] (4) is to be read as, if *p* were true, S would believe that *p*. Nozick provides a quick summary of the main idea: "To know that *p* is to be someone who would believe it if it were true [adherence], and who wouldn't believe it if it were false [variance]" (Nozick 1981, 178).

Nozick often appeals to possible worlds to clarify (3) and (4). Recall from Chapter 2 (§B) that Goldman deems the implications of Nozick's subjunctive theory for an account of relevant alternatives unsatisfactory. The problem was illustrated by the thermometer case, where the parent just happens to choose an accurate thermometer among several inaccurate ones. Assume, then, that the child's temperature is 98.6 degrees, and that the thermometer correctly registers this. In the *nearest* possible world where the child's temperature is not 98.6 – a world where the only difference is the child's temperature – the parent does not believe it is. If this is the *only* world relevant to the assessment of the subjunctive conditional (3), then it would seem that the sensitivity condition is satisfied. Goldman argues that this is the wrong result, for the parent only very luckily picked a good thermometer.

The present issue, then, is whether Nozick's interpretation of his subjunctive conditionals entails the *nearest* not-*p* world account of relevant alternatives.[3] I suggested in the last chapter that, though Dretske's and Nozick's sensitivity conditions are very similar, Nozick's may allow a broader spectrum of worlds to count as relevant. On Dretske's view, we must hold fixed all the circumstances in the particular, actual situation that are logically and causally independent of *p*.[4] If *p* is "the child's temperature is 98.6 degrees," then the thermometer used is logically and causally independent of *p*, and so must remain fixed when assessing what would be the case if *p* were false. If that's right, then, because the parent's belief satisfies Dretskean sensitivity, the parent knows the child's temperature. So if Goldman has a legitimate complaint against the permissiveness of the sensitivity condition in the subjunctive conditional view, it appears to apply to Dretske's account. Does it also apply to Nozick's sensitivity condition (3)?

Before trying to answer that question, I want to remind the reader that, even if Goldman is correct about the scope of relevant alternatives implied by Nozick's view, it is not entirely clear that Nozick's sensitivity condition (which again, I, unlike Nozick, restrict to his (3)), thus interpreted along Dretskean lines, yields the wrong answer in the thermometer case. The parent uses an accurate thermometer and forms a true belief, and in the nearest world where the proposition believed is false, she does not believe it. Perhaps that is enough for knowledge. It may not be the best or most solid kind of knowledge, but knowledge nonetheless. In Chapter 5, where I set out the main results of the dialectical process from Chapters 2 to 4, I will incorporate this thought, namely, that some examples hinge on intuitions that are just too unstable. Some cases receive widespread agreement – many Gettier cases, the Mary problem, the barn facsimiles example, and many others – and we should trust our intuitions. Where our intuitions are far less decisive in some cases, our theory should make room both for the possibility that such cases constitute knowledge and for the possibility that they do not. We should leave our options open, but have ready to hand an account on either side.

Nozick's "official" answer to the question of which worlds are relevant to the truth-value assessment of (3), and of subjunctives more generally, is found in a long footnote. (See note 8, 680–81; all quotations in this and the next paragraph are from this footnote.) He begins by saying that his purposes require "no more than an intuitive understanding of subjunctives," but things get messy very quickly. The first tentatively accepted interpretation of subjunctives that he gives is that a "subjunctive $p \rightarrow q$ is true if and only if *q* is true throughout the *p* neighborhood of the actual world." The *p* neighborhood is the band (Nozick's term) of worlds where *p* is true, such that there is no world where *p* is false that is closer to the actual world than that band. This reading of subjunctives is perhaps susceptible to Goldman's complaint that not enough counter-possibilities are deemed relevant in condition (3). Matters are obviously complicated, since (3) is a condition that requires belief to track the truth out to some relevant not-*p* world (rather than just through some band of *p* worlds), and so the application of the stated criterion of truth for counterfactual subjunctives is not direct. But the criterion does imply which not-*p* world is relevant, namely, the closest one. The closest not-*p* world in the thermometer case is where only the child's temperature is different, and everything else, including the thermometer used, is the same. Thus on this interpretation, if it were false that the child's temperature is 98.6, the parent would not believe it, and so the criterion would deliver a positive verdict on the parent's knowledge status, which generates Goldman's alleged counterexample.

But Nozick is not entirely satisfied with this account:

> This view of subjunctives . . . is inadequate if there is no discrete *p* band of the actual world, as when for each positive distance from the actual world A, there are both *p* worlds and not-*p* worlds so distant. Even if this last is not generally so, *many* p *worlds that interest us may have their distances from A matched by not-*p *worlds.* Therefore, let us redefine the relevant *p* band as the closest spread of *p* worlds such that there is no not-*p* world intermediate in distance from A to two *p* worlds in the spread unless there is also another *p* world in the spread the very same distance from A.
>
> (My emphasis)

I confess that it is not entirely clear why the exception requires that there be a *p* world the same distance from the actual world as the not-*p* world in question, but I think that the intuition Nozick is trying to capture is fairly straightforward and is applicable to the thermometer case. In the actual world, the child's temperature is 98.6 degrees and the parent uses the accurate thermometer T_a. In the nearest world where it is false that the child's temperature is 98.6, the parent uses the accurate thermometer T_a. But there is another close possible world where the child's temperature is 98.6 but the parent uses an inaccurate thermometer T_i. In both non-actual possible worlds, only one thing is different than in the actual world, and indeed, given that it is sheer chance that the parent selects the accurate thermometer in the actual world, it is natural for us to think that she easily could have selected a different one – that such a world is very (and relevantly) close to the actual world.

On any proposal for assessing the subjunctives that interest us, one's belief must track the truth through some relevant band of *p* worlds. Our question is: Does that band extend only to the nearest not-*p* world, or may it *include* some not-*p* worlds? Nozick seems to want to allow the second interpretation, and my suggestion is that the thermometer case illustrates why. If the world where the parent selects an inaccurate thermometer is deemed extremely close to the actual world – so close that, we want to say, it was in fact likelier that she did than did not, and everything else is the same in that world – then we should also count that world as relevant. (I *believe* this is an example that conforms to the exception criterion expressed in the quotation above.)

Thus a correct assessment of the subjunctive – if it were not the case that the child's temperature is 98.6, the parent would not believe that it is – *may* require us to consider as relevant some possible worlds that are (slightly) more distant from the actual world than the closest not-*p* world. In short, on a natural reading of subjunctives, it may be *false* that if it were not the case that the child's temperature is 98.6, the parent would not believe that it is. (Think about it this way. If you were to present this exact case to the person in the street and ask whether the subjunctive is true, you might very well be confronted with the reply: "I guess it depends on which thermometer the parent takes from the cabinet.") I will not go so far as to say that this is the definitive interpretation of Nozick's subjunctive account, but it seems like a plausible alternative to Goldman's *pure* subjunctive reading, one I believe Nozick was striving toward.[5] In any case, I am thus offering two distinct replies to Goldman's thermometer counterexample to the subjunctive conditional, or sensitivity view: (1) the parent does know, though she could know "better"; and (2) the parent does not know because she does not in fact satisfy Nozick's sensitivity condition.

The attentive reader might have noticed that all of this seems somewhat beside the point, since Nozick's fourth condition appears to be violated in the thermometer case: If *p* were true, S would believe that *p*. Surely, and as already noted, there is a very close possible world where it is true that the child's temperature is 98.6 degrees, but the parent does not believe it because she uses an inaccurate thermometer. The reason I have not exploited this feature of Nozick's account is because, for reasons we shall see, I do not accept Nozick's condition (4).

Let's take a brief test spin of Nozick's theory and see how it handles rough terrain.

Standard Gettier Cases. On Nozick's account, Smith does not know that either Jones owns a Ford or Brown is in Barcelona because, if this disjunction were false, that is, if Brown were somewhere else, Smith would still believe Brown is in Barcelona. Smith's belief violates Nozick's sensitivity condition (3) (Nozick 1981, 173).

Barn Facsimile. Nozick's condition (3) is violated here, too. If Henry's belief that the object he sees is a barn were false (because it is a facsimile), he would believe it anyway (Nozick 1981, 175).

Respect for Skepticism. Nozick is concerned not to merely dismiss skeptical possibilities as irrelevant to the question whether we can obtain knowledge: "An account of knowledge should illuminate skeptical arguments and show wherein lies their force. If the account leads us to reject these arguments, this had better not happen too easily or too glibly" (Nozick 1981, 197). S believes that she is not a BIV, but does she know it? This is a controversial issue, and in the next chapter we will explore a theory that is compatible with such knowledge. However, the skeptic's contention that we could not know that skeptical hypotheses are false is so strong that it has set the standard for epistemological theorizing for centuries. Nozick's theory respects this much of the skeptical viewpoint, in that if it were false that I am not a BIV, I would still believe I am not,[6] and thus my belief violates sensitivity; hence I do not know I am not a BIV. Because, however, Nozick's theory implies that knowledge of ordinary propositions is possible, and we all know that such propositions entail the negations of skeptical hypotheses, the theory commits him to denying that knowledge is closed under known entailment. (See Nozick's Chapter 3, §II.) This issue arose earlier in the introduction and in the discussion of Goldman's reliabilism. In Chapter 5, I shall outline a theory that accepts this consequence, and in Chapter 6 I shall try to explain away its counterintuitiveness.

Hologram Vase. S sees a hologram of a vase, which happens to be situated in front of a real vase, and so has a justified true belief that there is a vase in front of her. She does not know this, however, because if it were false, she would still believe it, based on her perception of the hologram. Condition (3) is violated.

The Lottery (LOTTERY). As Dretske (1971, 3–4) points out, sensitivity can be used to solve the lottery problem.[7] Suppose S has a one in a million chance of winning the lottery, and thereby comes to believe that she will lose. Probabilistically speaking, her belief is reliably formed – such statistical reasoning is truth-conducive in the actual world and throughout nearby possible worlds. Still, surely S does not *know* that she will lose. (If she knew it, why did she buy a ticket? S may be given to *saying* that she knows she will lose, and her belief seems justified. Even so, the chance that her ticket wins is identical to the chance that any other ticket wins, and thus it is implausible that she knows she will lose.) Even modalized versions of process reliabilism are hard pressed to explain lack of knowledge in the lottery case. But if S's belief that she will lose were false, she would still think she will lose, and that is why she does not know. (This assumes that we hold constant her method of belief formation. We of course ought not to consider her reading the lottery results in the newspaper when assessing the subjunctive conditional. See §B, below.)

Brain Lesion. S's brain lesion causes her to believe she has a brain lesion, and if it were false, she would not believe it. Still, it seems she does not know that she has a brain lesion.[8] But her belief violates condition (4) – she does not believe she has a brain lesion in nearby worlds where it is true, because in many of those worlds brain lesions do not cause brain lesion beliefs.

Necessary Truths (NECESSARY TRUTH). Logical truths are not possibly false, and so Nozick's condition (3) seems ill-suited to explaining how we track them – its antecedent is never satisfied. However, when S *knows* some logical truth, (4) is satisfied. She believes it not only in the actual world but also in many close possible worlds. If we suppose that S merely guessed that some logical truth holds, her belief violates (4) because there are close worlds where she does not believe it, and thus she does not know. Nozick therefore claims that only conditions (1), (2), and (4) are required for knowledge of necessary truths (Nozick 1981, 186f.).

Tom Grabit (GRABIT). S sees Tom Grabit steal a book from the library, and thus forms the true belief that Tom Grabit took it. Suppose that, unbeknownst to S, Grabit's mother says that Tom is out of town, but she is lying.[9] Does S know that Tom took the book? One intuition is that the answer is "no," because if S had heard what Grabit's mother said, she would have believed it. This violates Nozick's condition (4) – there is in fact a close world where it is true but S does not believe it because S hears what Grabit's mother said.[10]

Though not exhaustive of cases Nozick examines, the menu above offers a nice sampling of ways in which the subjunctive conditions can be violated when an agent lacks knowledge. I shall presently attend to the aforementioned important refinement concerning methods in Nozick's theory, but want now to indicate briefly reasons why, in the end, I will reject Nozick's condition (4). In short, (4) is both too restrictive and, more importantly, too permissive. One of its first uses is to answer the DICTATOR problem raised by Harman.[11] Suppose a dictator of a country is killed, and it is accurately reported in the early edition of a major newspaper. Later all the newspapers deny the story, and everyone who reads one of them believes the denial, or at least suspends judgment. However, S does not read any of the denials. Does S know that the dictator was killed? Nozick follows Harman in denying that S knows, and his condition (4) can be used to explain why – there is a nearby world where it is true that the dictator was killed but S does not believe it, a world where she reads the later papers. This is analogous to the Grabit case; they both include a possible defeater to one's actual grounds for believing. On the other hand, it is not entirely clear that Nozick's and Harman's intuitions are correct. After all, in both cases one's grounds for belief are very strong. Perhaps this is clearer in the Grabit case than in the dictator case; perhaps Grabit does know. If so, notice that if we supplement sensitivity with modalized process reliabilism while discarding (4), we get some positive results. S (in fact) knows that Grabit took the book because (a) if it were false, she would not believe it; and (b) S's vision is reliable in the actual world and throughout nearby worlds. The dictator case is admittedly more controversial. S perhaps does *not* know that the dictator was killed because, even though (a) if it were false, she would not believe it; (b) her belief-forming process – reading the paper – is not truth-conducive through close possible worlds. At least, it is not if the papers are prone to lie about important political matters.[12] In these cases, the difference in knowledge status between the agents hinges on whether their beliefs are formed by a reliable process.

My suggestion so far is that (4) is too restrictive, improperly ruling out cases of knowledge, such as Grabit. (As will become apparent in the discussion of methods of belief formation, Nozick's proposal to hold the method constant in assessing the subjunctive conditions is inconsistent with his diagnosis of the Grabit and dictator cases, where (4) is violated only if methods are *not* held constant.) That (4) is too restrictive becomes even more plausible if one takes seriously the suggestion that some cases of knowledge are better than others, which allows for instances of knowledge that may not be all that we would want, but which appear to be genuine anyway. We have also seen, however, that there are other cases where (4) seems to be a reasonable addition to Nozick's epistemology, such as the brain lesion. Here again, modalized process reliabilism may be a reasonable replacement for (4). The victim of a brain lesion that causes true brain lesion beliefs does not have knowledge if the cause is well-nigh miraculous, but if such a lesion were to cause true brain lesion beliefs in the actual world and throughout nearby worlds (which, by hypothesis, it does not), such beliefs might very well constitute knowledge.

Assuming, then, that it is beneficial to substitute a process reliabilist requirement in place of (4) in the cases just discussed, we are left with necessary truths as the strongest rationale for the addition of Nozick's (4) to the theory. But even here (4) faces problems, as Garrett demonstrates. Suppose that a mathematician believes FERMAT'S LAST THEOREM to be true because no counterexample has been discovered, and suppose his belief is true. His belief then satisfies (4), for he believes it in all nearby worlds. "But [he] does not know that Fermat's Last Theorem is true. His belief to that effect was reached via an unreliable method without the aid of any reliable sub-method."[13] (4) is too permissive.

Just as Garrett suggests, here again modalized process reliabilism can be coupled with sensitivity to achieve the desired result. If one believes in the law of noncontradiction because she likes the sound of it and stubbornly clings to her belief (or whatever), it will not violate sensitivity because there is no world where it is false, but the process is extremely unreliable, yielding many false beliefs. In the following discussion of methods of belief formation, we will find more reasons still for rejecting (4), leaving us to consider whether sensitivity (3) is a necessary condition for knowledge. As you already know, I think it is, but fuller elaboration of the sensitivity plus modalized process reliabilism view will have to wait until Chapter 5.

B. Methods: An important refinement

In this section, I work to solidify the conclusion that Nozick's (4) should be rejected in favor of a process reliability condition. The issues here are very complex, so as far as possible, I try to stick to considerations bearing on my main conclusion, leaving others to endnotes. Having just argued that (4) is both too restrictive and too permissive (where the latter is especially important), I explain here how Nozick needs to relativize his subjunctive conditionals to the agent's specific method of belief formation in order to accommodate plausible cases of knowledge where one or the other of his tracking conditions would otherwise be violated. We will see, however, that Nozick inconsistently applies his conditions – sometimes holding an agent's method of forming belief constant, sometimes allowing it to change. I shall argue that it is the need to make Nozick's condition (4) fit our judgments about agents' knowledge statuses in various cases that forces the inconsistency, and that this is further reason to reject (4). Discarding that condition, we can consistently retain the sensitivity condition (3) in our theory, but, again, we'll need to supplement it with a modalized process reliability condition to take up the slack created by the absence of (4). Here I shall also try to clarify the issue of method individuation.

Nozick offers a convincing case (GRANDMOTHER) for why his subjunctive conditionals must be relativized to one's method of forming a belief. "A grandmother sees her grandson is well when he comes to visit; but if he were sick or dead, others would tell her he was well to spare her upset. Yet this does not mean she doesn't know he is well when she sees him" (Nozick 1981, 179). If he were unwell, his grandmother would believe he is well anyway; thus her belief apparently violates sensitivity.

Here are Nozick's plausible amendments to (3) and (4):

(3) If *p* weren't true and S were to use [method] M to arrive at a belief whether (or not) *p*, then S wouldn't believe, via M, that *p*.
(4) If *p* were true and S were to use M to arrive at a belief whether (or not) *p*, then S would believe, via M, that *p*. (*ibid.*)[14]

If her grandson were ill and she were to use the same method of belief formation as she actually does, she would not believe that he is well, because she would *see* that he isn't. When we assess sensitivity relative to the method actually used, the grandmother's belief does not violate that condition. This example makes clear why relativization to a method of belief formation is plausible, and it provides a nice illustration of different methods, in that vision and testimony (from her family) are obviously distinct. It does not, however, indicate any clear criterion for method individuation. Consider a case from Goldman (1983, 84; DACH). "Oscar sees Dach the dachshund and believes there is a dog before him. If there weren't a dog before him, a hyena would be there instead, which Oscar would misclassify as a dog." Does Oscar know that there is a dog before him? If it were false, he would believe it, but for all that, it seems he knows that there is a dog before him anyway. Jonathan Vogel describes how Nozick could distinguish Oscar's methods to account for his knowing that what he sees is a dog: "One way of characterizing Oscar's actual method would be something like 'If you seem to see a small, squat, elongated animal with short legs and floppy ears, that barks, etc., conclude that there is a dog before you'."[15] Were Oscar to apply his actual method in the counterfactual circumstance where the thing seen is a hyena, he would not form the belief that it is a dog.

These two examples suggest very different criteria of method individuation – one very broad (vision versus testimony) and one very narrow (inference from floppy-eared-looking thing versus inference from spotted-looking, shrill-sounding thing). Nozick seems to prefer a narrower interpretation:

> A person can use a method (in my sense) without proceeding methodically, and without knowledge or awareness of what method he is using. Usually, a method will have a final upshot in experience on which the belief is based, such as visual experience, and then (a) no method without this upshot is the same method, and (b) any method experientially the same, the same "from the inside", will count as the same method.
>
> (Nozick 1981, 184–5)

Echoing this idea of how methods are individuated, Nozick says, "The method used must be specified as having a certain generality if it is to play the appropriate role in subjunctives. This generality is set by the differences the person would notice; the methods are individuated from the inside" (Nozick 1981, 233).[16] These passages suggest that Nozick favors a very narrow individuation of methods – "no method without this upshot is the same method" – and we have seen how this can help in the Dach case. And yet, as Nozick says, the method must have some generality if we are to make sense of the subjunctives. It is not entirely clear what he means by this, but at a minimum it would seem that the same method must be applicable to different possible experiences. (It seems that Nozick does not always honor this condition. See note 18.)

Some commentators have pointed out that indexing the subjunctive conditionals to the method actually used is inconsistent with Nozick's treatment of several examples, and thus that no univocal view of method individuation is consistent with plausible answers to some reasonable test cases. Robert Shope, for one, has shown this.[17] Consider first the JESSE JAMES case, where S is a bystander who happens to see an escaping robber's mask slip off and realizes that it is Jesse James (Nozick 1981, 193; Shope 1984, 35). Though it is correct to say that S knows it is Jesse James, S's belief violates Nozick's condition (4) – there is a very close world where it is Jesse James but S does not believe it because the mask does not slip. Nozick attempts to solve this problem by appeal to a very specific method: "His method is looking and concluding it is Jesse James on the basis of seeing certain things. By that method (applied in this way) he would know in other cases also. To be sure, some other situations might not allow that method to be used … but that causes no difficulty for condition 4, properly understood" (Nozick 1981, 193). In the closest worlds where it is Jesse James and S uses that specific method, one that holds constant *what she saw*, she believes it is Jesse James. The suggestion seems to be that, in the nearby world where the mask does not slip, S does not use the method described because S has different experiences, and thus there is *no* nearby world where she both *uses that* method and does not believe it is Jesse James when it is.[18]

Shope shows how individuating methods by holding constant specific experiential content is inconsistent with Nozick's reply to the dictator case. S believes truly that the dictator was killed, but would have believed falsely that he was not killed had he read the later newspapers. Thus Nozick says that S does not know because "if he had heard the denials, he too would have believed them" (Nozick 1981, 177). If experience must be held constant, however, this conditional appears improper. S's method would be something like *believing on the basis of reading about the dictator's death in the morning paper*. Shope says, however, that "if the content of what is read and heard is fixed in the manner we supposed regarding [the Jesse James case], then it is irrelevant for Nozick to point out the happenstance that certain items did not actually figure in that content" (Shope 1984, 37). Whereas in the Jesse James case, Nozick claims that what is seen must be held fixed for application of the same method in nearby worlds, his reply to the dictator case conflicts with that claim, for "if the content of what is read and heard is fixed in the manner we supposed regarding" the Jesse James case, then S would believe (truly) that the dictator was killed, which undermines the intuition that S does not know that he was killed. Now, I have already claimed that it is not obvious that S does not know this, but that is not the basis of the objection here. Rather, the central point is that Nozick inconsistently applies his conditions, sometimes holding methods fixed and sometimes not.

Nozick's replies to the dictator and Grabit cases are *clearly wrong*, for they undermine his solution to the grandmother case. In the dictator and Grabit cases, Nozick attributes lack of knowledge to violation of (4), but only by failing to hold the methods fixed. (S might have read something different in a later paper; S might have heard what Grabit's mother said.) Applying Nozick's reasoning about the dictator and Grabit examples to the grandmother case, one could say (wrongly) that grandmother does *not* know because if her grandson were ill she would have been told he was fine and believed it.

We ought, then, to index the application of Nozick's conditions to the method actually used, and to individuate methods narrowly, by reference to actual experience, for it is on that basis that one forms specific beliefs. We can then conclude, *pace* Nozick, that S knows that Grabit stole the book (which I think is true) because her belief, relativized to her actual method, satisfies both (3) and (4). But we still have to contend with the fact that (4) is too permissive for necessary truths. We can solve *that* problem by replacing (4) with a process reliability requirement. The solution generalizes to the dictator, Grabit, and Jesse James cases. S knows if and only if her true belief is sensitive and formed by a reliable process. The Grabit and James cases thus constitute knowledge, whereas the dictator case is (predictably) less clear because whether one knows that he was killed depends on whether the process of forming beliefs based on reading the paper is deemed reliable, and that is a vexed question. We no longer need (4), which is a positive result, given the inconsistency in the way Nozick wields it.[19]

My proposal, then, is that we replace Nozick's condition (4) with a process reliability requirement, and individuate methods narrowly, in particular, for empirical beliefs,[20] by reference to one's actual experience and belief content (believe that *p* if you have experiences of type ϕ), but not by requiring that actual experience itself be held constant for counterfactual applicability of the same method. (See note 18.) If it were false that there is a dog before Oscar, he would not believe that it is a dog based on his method of concluding it is a dog given certain specific appearances (because those appearances would not obtain). If it were false that the grandson is well, the grandmother would not believe he is well because she would not have the visual experience as of a healthy person. If it were false that the robber is Jesse James, S would not believe that it is, using her method of concluding who it is on the basis of seeing the person's face. (If it were false that the dictator was killed, S would still believe he was based on reading the morning paper. Or perhaps S does know that the dictator was killed because if it were false the morning paper would not have said he was killed. No wonder that this case garners contradictory intuitions.) In all these cases, the competing demands to make condition (4) come out right cause Nozick to apply it inconsistently. Abandon (4), and our problems are solved. Testing the hypothesis requires looking at cases where sensitivity, condition (3), alone is too weak, such that some further condition is required. I turn to such cases in the next section, and after that, to cases where sensitivity may be too strong, ruling out genuine instances of knowledge.

Before turning to those objections to Nozick's theory, we ought to consider a further wrinkle caused by the relativization to method. Suppose an agent believes that *p* based on two distinct methods – her belief is overdetermined. Nozick considers a case where a father believes on distinct grounds that his son is not guilty of a crime (FAITHFUL FATHER): (1) because the jury acquitted him; and (2) because of his faith in his son (Nozick 1981, 182). The first issue is whether the extremely fine criterion of method individuation can even allow us to distinguish these methods – why not call it a single, very specific method? We can say that they are different methods precisely because each of them, independently of the other, would yield a belief as to whether the son is innocent. (I shall assume that there is a convenient way to incorporate this aspect of method individuation into the criterion.)

Second, how do we determine whether the father knows that his son is innocent? Which is the relevant method to hold constant when assessing the subjunctive? These questions become pressing when we imagine that, were the son guilty, the father's faith in his son would lead him to believe that he is innocent anyway, whereas the jury's conviction would lead him to believe otherwise. We should follow Nozick in claiming that the father knows only if the method based on the jury's conviction "outweighs" his faith, that is, if in the possible world where the son is guilty, the father would accept (the method based on) the jury's verdict and give up (the method based) on faith. Thus, more generally, one knows that *p* only if, when one's belief that *p* is overdetermined by methods, were the proposition believed false, the method that outweighs all others would lead one not to believe it.

C. Objections to Nozick's theory

1. General objections

One very general objection to Nozick's theory is that it claims to achieve some anti-skeptical results but does not. Edward Craig, describing Nozick's view, writes: "[I]n assessing the counterfactuals implicit in [ordinary knowledge claims] we do not have to take sceptical worlds into the reckoning, whereas we must do that in assessing [claims to know the denials of skeptical hypotheses] because [the latter] explicitly speaks of them."[21] Craig claims that one can ignore skeptical possibilities when assessing whether ordinary knowledge claims satisfy Nozick's conditionals only by *assuming* that the actual world is not a skeptical world, which achieves nothing as a reply to the skeptic.

One can easily understand Craig's misgiving here. But Nozick does not attempt to show that we actually know many things, only that it is possible to know ordinary truths without knowing the denials of skeptical hypotheses that they entail. Thus *if* (some everyday proposition) *p* is true and S believes it, S knows only if her belief tracks the truth throughout nearby worlds centered around the actual *p* world. If the actual world is one where some radical skeptical hypothesis is true, one does not know that *p* because it is false. But the skeptic aimed to show that lacking the ability to distinguish ordinary worlds from skeptical ones is a reason for saying we simply *cannot* know everyday truths. Nozick aims to block this skeptical strategy, which does not require that *we are in a position to say* that the actual world is not, for example, a BIV world. Thus on Nozick's view, if in fact it is not, then we may very well know lots of things.

As Anthony Brueckner rightly points out in his reply to Craig, Nozick's theory entails the denial of closure.[22] Avoiding the obligation to show that we are not in a skeptical world requires arguing that lacking such knowledge is irrelevant to the possibility of knowing ordinary propositions. It must be the case that it is possible to know that *p* (I have hands), that *p* entails *q* (I am not a BIV), and yet fail to know that *q*, and this is made possible by the claim that the nearest worlds relevant to whether one knows that *q* are not relevant to whether one knows that *p*. Unless Nozick denies closure, which he does, the question whether one knows that *q* will be directly relevant to the question whether one knows that *p*. On the other hand, the denial of closure is counterintuitive. We have already added this issue to our list of topics that require further treatment once we have set out our theory in Chapter 5.

2. Objections: sensitivity is too permissive

In this section, the need for including a process reliabilism condition becomes even clearer. We will also look more at the contrast between the narrow individuation of methods (in Nozick's sense) and the broad individuation of processes (and methods, in Goldman's sense).

Graeme Forbes (1984, 45) points to the counterintuitive result that sensitivity yields in the following case from Nozick (VASE-TRACKING HOLOGRAM).

> A person comes to believe a vase is in a box by seeing an illuminated hologram, part of a machine that alternates between displaying the hologram and the real vase contained in the box. . . . [T]he machine, in alternate time periods, displays a hologram of a vase only when a vase is pressing down on a lever. . . . Hence if there weren't a vase there, he wouldn't believe there was one.
>
> (Nozick 1981, 190)

The person's belief satisfies sensitivity, but apparently he does not know that there is a vase in front of him.

Alan Goldman has us consider "a standard Gettier case . . . I have been tricked into believing that someone in my office owns a Ford by Haveit's giving me evidence that Nogot owns one, when Haveit is the owner. If my belief were not true, then Haveit would not have led me to hold it. Yet I do not know that someone in the office owns a Ford" (HAVEIT).[23] Here again the belief satisfies sensitivity but does not constitute knowledge.

These are cases where once again it may be helpful to include a global reliability requirement – the belief-forming process must yield a high ratio of true beliefs in the actual world and in nearby worlds. Processes (in Alvin Goldman's sense, which for present purposes here *include* what he calls methods) are not individuated as finely as methods (in Nozick's sense). The process in the Haveit case is something like belief based on testimonial evidence, but that, like vision, is entirely *too* broad. Generally speaking, vision and testimony are truth-conducive belief-forming processes, but our aim is to locate the unreliable (sub)-processes at work in the vase-tracking hologram and Haveit cases. More narrowly individuating the belief-forming process based on vision is necessary to diagnose the lack of knowledge in the vase-tracking hologram case; there is in fact *something* clearly unreliable about beliefs based on hologram sightings, and any such true beliefs therefore involve process luck. Similarly, we must be able to narrow, without ad hoc stipulation, the testimonial process in the Haveit case in order to conclude that the agent's belief that someone owns a Ford is unreliably formed. Perhaps we can appeal to the fact that Haveit's testimony is motivated by his desire to dupe the believer, in which case beliefs based on what Haveit says also involve process luck because that process is unreliable. Settling on a principled means of individuating processes is an issue I shall pursue when we return to the generality problem in Chapter 5, after outlining the theory I prefer. For now, I want to say that the agent does not know that someone owns a Ford in the Haveit case because, though his belief satisfies sensitivity, it is not reliably formed.[24]

Luper-Foy poses a similar problem case (1984, 31, FLIP). Suppose Flip bases his beliefs about whether the light is on in the next room on a coin toss. Suppose further, however, that Jessica is watching Flip, and when the coin comes up heads, turns the light on, and when tails turns it off. So if it were false that the light is on, Flip would not believe it. But surely Flip does not know that the light is on when it is. Here again, adding a global reliability requirement can help. The process Flip uses is basing beliefs on a coin toss, and that general process is utterly unreliable, which is why Flip does not know.

3. Objections: Sensitivity is too restrictive

More difficult objections aim to show that Nozick's sensitivity requirement is too restrictive. After all, if it rules out cases of genuine knowledge, there is nothing we can *add* to the theory to get the desired results. I shall focus on two commonly highlighted worries.

First, Nozick's analysis makes it difficult to account for inductive knowledge. Here is an example (TRASH CHUTE) from Sosa. "On my way to the elevator I release a trash bag down the chute from my high rise condo. Presumably I know my bag will soon be in the basement."[25] But if it were false that it will soon be in the basement, I would believe that it will soon be there anyway. I know, but my belief fails to satisfy sensitivity. Here is another similar case (THE ROOKIE) from Vogel. Two policemen confront a mugger who is some distance from them. One of them, a rookie, tries to disarm the mugger by shooting a bullet down the barrel of the mugger's gun. The rookie's partner knows he is trying to pull off this all but impossible feat, and believes that the rookie missed. If it were false, he would still believe it using his actual method, but surely he knows that the rookie missed anyway (1987, 212).

The key element to both of these examples is that there is a very unlikely possibility that is relevant to the assessment of the sensitivity conditional. Because of the improbability, it seems right to say that the agents have knowledge, even though their beliefs violate sensitivity. Nozick's reply to this kind of case appears to be inconsistent with his general claim that the method must be fixed. It seems right to say, for example, that we know that the sun is going to rise tomorrow, but if it were false, we would believe that the sun will come up anyway (SUNRISE). Nozick replies, "If the sun were not going to rise tomorrow, would we have seen that coming, would that alteration in the earth's rotation have been presaged in the facts available to us today and before?" (1981, 222) If we must hold the method fixed, it is something like basing one's belief about whether the sun will come up on evidence that it has every day in the past. If it were false that the sun will come up, one would still believe it true according to this method. To solve this problem, one where sensitivity violation appears to rule out genuine knowledge, Nozick wonders whether the method used would be different in the counterfactual scenarios, such that, were the sun not to rise, one would not believe that it will, based on the alternative method. But that is to allow methods to *change* in assessing the counterfactuals. Shope rightly says that "holding the method fixed [as Nozick's demands require in other cases] does not allow us to consider inferences from other beliefs that we might have" (1984, 34).

The particular counterexample construction method here is ingenious. Pick a case of belief where the relevant not-*p* possibility is so unlikely that it is easy to violate sensitivity and yet, to that same extent, plausible to say that S knows that *p*. Our dilemma is to give up sensitivity or to claim that some of the securer cases of induction do not constitute knowledge that an event will happen.

If we take the dilemma as inescapable, which we need not (see below), we might well deny knowledge of the proposition in question in each of these cases *while at the same time accounting for inductive knowledge*. We can plausibly say this: S does not know that the trash bag will soon be in the basement; rather, S knows it is *very likely* that it will soon be there. The policeman does not know that the rookie missed, but that it is *very likely* that the rookie missed. This is not to invoke Humean skepticism about induction. That kind of skepticism hinges on the supposed groundlessness of the principle that the future will be like the past. But here we are allowing that kind of induction principle, and thus our denying that S knows is not simply based on skepticism about the principle. If it is possible that the trash gets stuck or the bullet lodges in the mugger's gun, then, without further (after the fact) evidence, S knows only that it is extremely likely that the trash will be in the basement, and the veteran cop knows only that it is very likely that the rookie missed. This may seem like a drastic claim, but it is supported even by introductory logic textbooks, almost all of which define an inductive argument as one that makes the conclusion *probably* true. That, I am suggesting, is reason enough to deny that S knows that the trash will soon be in the basement, and that the policeman knows that the rookie missed, while not precluding inductive knowledge.

Some will nonetheless think that denying knowledge that some specific event will occur in these cases goes too far, perhaps especially in the sunrise case. To mollify those complaints, we might also try to reject the dilemma (that sensitivity is false or inductive knowledge does not exist) altogether.

To do so, we need to clarify further the methods of belief formation. We said that S believes the sun will rise based on the fact that she has never experienced a day when that did not happen. Perhaps we should add: And there is no evidence that it will not. Surely if there were such evidence, S would use it. (This is the basis of Nozick's reply.) In general, it is of the very nature of inductive inference that the method includes not only accounting for all the positive evidence, but also at least implicitly considering possible counterevidence. S's method, then, is to infer whether the sun will rise tomorrow based on all the evidence, positive and negative. If the sun were not to rise tomorrow, then, using *that* method, S would not believe that it will (assuming that evidence of alteration in the earth's rotation would have been "presaged in the facts," as Nozick says). In the trash chute case, S's method is to infer that the trash will go to the basement based on all the evidence. If the trash bag were not to fall to the basement, S would not believe it will if there is evidence that the chute is somehow clogged. Now, if S would not notice or obtain evidence that the chute is clogged if it were, that is reason for reverting to the assessment of the previous paragraph – S knows at best that it is *likely* that the trash bag will drop to the basement.

The upshot of these reflections is that the sensitivity requirement is not inconsistent with inductive knowledge. Theorists sometimes take inductive knowledge claims too literally. Often one does not know that an event will occur, but only that it will very likely occur. In addition, if we sharpen the inductive belief-forming method to include evidence of both positive and negative instances, we can even make sense of an agent having inductive knowledge that an event will occur.[26] If the agent is receptive to possible counterevidence, then her method yields sensitive beliefs. If not, the agent at best knows that some outcome is likely.[27]

The second argument that sensitivity is too restrictive focuses on higher-level beliefs, such as an agent's belief that her belief is not false. This line of argument is very influential and has led many to believe that the sensitivity requirement is far too strong in that it is impossible to satisfy in rather ordinary cases of higher-level knowledge. I should like, therefore, to treat this problem at some length. Sosa (1999, 145) offers a version of this criticism, but for present purposes, we will focus on Vogel's similar but more exacting argument.[28] Vogel argues that sensitivity cannot make room for knowledge that one's belief that *p* is *not false*, from which it follows, he claims, "that you also fail to know that your belief is true," hence that sensitivity "makes it impossible for you to know that *any* of your beliefs is true, not false" (Vogel 2000, 611). Interestingly, there is a parallel argument, based on an alternative but, I believe, more plausible interpretation of the main premise in Vogel's argument, which squares sensitivity with higher-level knowledge both that one's belief is true and that one's belief is not false. It follows that, while Vogel's argument reveals the incompatibility of sensitivity and knowledge of certain higher-level propositions, it does not establish the general claim that sensitivity is incompatible with knowledge that any of one's beliefs is true, or not false. Thus even though Vogel's argument raises a prima facie problem for sensitivity, things are not as bad for the sensitivity theorist as originally advertised. Indeed, I don't believe things are bad at all, precisely because Vogel's crucial premise hinges on a translation into a logical symbolism that is a prima facie *incorrect* rendering of its natural language higher-level belief counterpart. If I am right that Vogel incorrectly construes the relevant higher-level beliefs in aiming to show that sensitivity and higher-level knowledge are incompatible, and that otherwise sensitivity is compatible with higher-level knowledge, then Vogel's argument will be completely defanged.[29]

Here is Vogel's (2000, §II) argument that sensitivity fails to allow for higher-level knowledge. (I shall use Vogel's original proposition numbering.) Suppose you meet a friend, Omar, and notice that he has new shoes, and he tells you that they are new (OMAR). The following are true:

(10) You know Omar has a new pair of shoes.
(11) You know that your belief that Omar has a new pair of shoes is true, or at least not false.

Vogel argues that sensitivity cannot accommodate (11), though he focuses on knowing that your belief is not false, rather than on its being true. Here I shall quote at length.

> Let O = 'Omar has new shoes'; B(O) = 'You believe Omar has new shoes'; B(¬(B(O) & ¬O)) = 'You believe that you do not believe falsely that Omar has new shoes'.
> (14) In the actual world, B(¬(B(O) & ¬O)).
> Consider the nearest possible world W in which your higher-level belief is false, that is,
> (15) In W, ¬(¬(B(O) & ¬O)).
> Simplifying,
> (16) In W, (B(O) & ¬O).
> From (16),
> (17) In W, B(O).
> If you believe O, you believe that you do not falsely believe O ... Hence,
> (18) In W, B(¬(B(O) & ¬O)).
> Given (15) and (18),
> (19) In W, (B(O) & ¬O) & B(¬(B(O) & ¬O)).
> That is, if your belief that O were false, you nevertheless would believe that it was not false. So, according to [sensitivity], you fail to know that you do not believe falsely that O. It follows, I take it, that you also fail to know that your belief is true.
>
> (Vogel 2000, 610–11)

Does it in fact follow that you fail to know that your belief is true? Let us repeat Vogel's argument with this possibility in mind.

The crucial step in Vogel's argument is (14), where he focuses on your actual world belief that *you do not believe falsely.* Note that Vogel's (14) does not entail that you have the belief that O, only that it is not the case both that you believe that O and that O is false. The significance of this will become clear later. Taking (14) as our starting point, the relevant counterfactual world specified by (15) is one where it is false that it is not the case that you believe that Omar has new shoes but Omar does not have new shoes – in short, it is false that you do not believe falsely. But we can recast (14) and (15) in terms of the question whether you can know that your belief is true:

(14*) In the actual world, B(B(O) & O).

You believe that your belief is true[30]; you believe that you believe that O and that O is true.

(15*) In W, ¬(B(O) & O).

Where in the actual world, you believe both that you believe that O and that O is true, in W this higher-level proposition is false. (Following Vogel, I shall use "W" to stand for the nearest world where the proposition believed is false.) It is not the case both that you believe that O and that O is true. From (15*) it follows that:

(16*) In W, ¬B(O) v ¬O.

This says that, in W where the higher-level proposition believed in the actual world is false, either you do not hold the first-order belief, or the first-order belief is false (or both). (Put in subjunctive terms: Were your actual world higher-level belief false, either you would not believe that O, or O would be false. I prefer this construal because it makes clear that you may very well not have your actual world higher-level belief in some possible world where the higher-level proposition is (or would be) false.) Now, if it is the case that the higher-level belief would be false because in W you do not hold the first-order belief that O, then in W you also would not believe that you believe that O.[31] In that case, your higher-level belief is truth-tracking: If your higher-level belief – that you have a (true) belief that O – were false, you would not hold that higher-level belief. On the other hand, if your higher-level belief is false in W because O is false, then, assuming as we have that your first-order belief is truth-tracking – that it is a case of knowledge – here again you would not believe that O, and presumably would not hold your actual world higher-level belief in W. In either case, it is certainly possible for your higher-level belief to constitute knowledge – to satisfy sensitivity.

If this is right, how can it conflict so radically with Vogel's result? How could it be that you cannot know that you do not believe falsely, but you can know that your belief is true? Well, it turns out that on a different interpretation of believing that your belief that O is not false, you *can* know that your belief is not false. Here's how that argument goes:

(14#) In the actual world, B(B(O) & ¬¬O).

You believe that you believe that O and that O is not false. The higher-level proposition in (14#) is false in W:

(15#) In W, ¬(B(O) & ¬¬O).

From (15#), (16#) follows:

(16#) In W, ¬B(O) v ¬¬¬O; that is:

(17#) In W, ¬B(O) v ¬O.

This is the result we achieved above, namely, that in the nearest world where your actual higher-level belief is false, either you would not believe that O (and hence not believe that you believe that O), or O would be false. As before, if the latter is the case, and given that you do know that O, if O were false, you would not believe it, and so would not believe that you believe it. So in either case, your higher-level belief is truth-tracking.

It appears, then, that, with respect to the question – can you know that your belief is not false? – the issue turns on the kind of higher-level belief in question: whether the actual world higher-level belief is characterized as it is in (14) or (14#). Now, I have admitted that Vogel is right that sensitivity is incompatible with the higher-level belief in (14), and that this *seems* problematic. However, it appears that the higher-level belief in (14) is not the best translation of the original target proposition:

(11) You know that your belief that Omar has a new pair of shoes is true, or at least not false.

To see this, consider again Vogel's translation of the higher-level belief implicit in this proposition:

(14) In the actual world, B(¬(B(O) & ¬O)).

This says, literally, "You believe that it is not true both that you believe that O and that O is false." This is indeed a natural way of saying, "You believe that it is not the case that you have a false belief," or "You believe that you do not believe falsely." What we want to symbolize is a version of (11): "You believe that your belief is not false," hence (14#):

(14#) In the actual world, B(B(O) & ¬¬O).

The former constructions say you believe that *you do not have* a belief that is false, the latter that you believe that *you have* a belief that is not false. Only the latter entails that you actually do believe that *p*, which is what a translation of (11) demands. From (14), Vogel construes the relevant W as (15), a world where you maintain your first-order belief, but it is false. ((16) shows this clearly.)

The upshot of all this is that one can have higher-level knowledge that one has a belief that is true and that one has a belief that is not false, but not that one does not have a belief that is false. When one construes the relevant higher-level belief in either of the first two ways, it turns out that sensitivity is violated precisely when one would expect – when one does not know. There are two possibilities: Either S knows that *p* or S does not know that *p*. If S knows that *p*, then, as we've seen, in the nearest worlds where *p* is false, S does not believe that *p*, and so does not believe that she believes that *p*. That is, she has no false higher-level belief. Suppose that S does not know that *p*, though *p* is true. Then in the nearest worlds where *p* is false, S may still believe that *p*, and falsely believe that her belief that *p* is true. But surely this lack of higher-level knowledge is not a problem for the sensitivity theorist, for if S does not know that *p*, it is not controversial to claim that she does not know that her belief that *p* is true.

Construing higher-level beliefs in Vogel's way fails to capture the fact that S has first-order knowledge, but it is only when we assume that an agent has first-order knowledge that we can expect higher-level knowledge (as discussion of the second disjunct, previous paragraph, shows). Since Vogel's translation of the relevant higher-level belief is compatible with S not even believing that *p* (14), *a fortiori* compatible with S not knowing that *p*, it's no wonder that there will be nearby worlds where S believes that *p*, hence believes that she believes that *p*, but *p* is false, as in (15) and (16). I conclude, then, that on a proper understanding of the relevant higher-level beliefs, sensitivity is compatible with higher-level knowledge.

D. Conclusion

Nozick constructed an ingenious theory to solve the Gettier problem and to achieve some anti-skeptical results while honoring a fundamental truth in skepticism. His subjunctive conditionals aim to capture the idea that knowledge is believing that *p* when it is true and not believing it when it is not. It follows that, while knowledge of ordinary propositions is certainly possible, we fail to know the denials of skeptical hypotheses.

Putting Nozick's analysis through the wringer, we have seen that his condition (4) causes deep difficulties. First, it is sometimes too restrictive. Nozick applies (4) to diagnose lack of knowledge in the Grabit case: There is a nearby world where Grabit took the book but S does not believe it because she heard Grabit's deceitful mother say that Tom Grabit is out of town. But surely it is not obvious that S does not know – she *saw* Tom take the book, knows what Tom looks like, and Tom has no twin. Second, this very response of Nozick's to the Grabit case is inconsistent with the need to fix the belief-forming method in assessing the subjunctives in other cases, such as the grandmother. If the method must be held constant, then it is right to say that S does know that Grabit took the book. Perhaps the correct upshot of this inconsistency is to reject Nozick's original solution to the Grabit case and accept that S knows that Grabit took the book, not, in my view, because S satisfies both Nozick's (3) and (4), but because S satisfies (3) and forms the belief from a reliable process. Then there was the problem that (4) is too permissive, allowing cases of knowledge of necessary truths where an agent does not in fact know them. Here again, replacing (4) with a modalized process reliability condition separates cases of knowledge from cases of lack of knowledge. Because (4) is too restrictive, too permissive, and is not consistently applied to different cases, we ought to reject it in favor of the requirement that one's belief be formed by a reliable process.

We have also considered a host of potential problems with sensitivity (Nozick's (3)). The first class of problems suggests that sensitivity is too weak, such that belief can vary with the truth in the required way, but the agent does not know. Haveit provides evidence that he owns a Ford only when it is true that someone in the office (but not Haveit) owns a Ford; Flip believes the light in the next room is on only when it is. Even the commentators who raise these problems claim, however, that the agents' beliefs are not reliably formed, and thereby implicitly refer to the problem of process luck. This supports the contention that including a process reliability condition, one that covers all the cases that prompt Nozick's inclusion of (4), but even better, is well motivated.

The second class of problems aims to show that sensitivity is too strong, for it is inconsistent with inductive and higher-level knowledge. With regard to induction, I offered two distinct responses. One is that the knowledge one has by induction is often too generously characterized. The policeman does not know that the rookie missed, but only that it is extremely likely that he did. The second response (for those who shudder at the first) is that the inductive method itself makes room for defeating evidence. In the sunrise case, let us assume that S knows the sun will rise tomorrow. Her method is not merely that she has noticed that it has always come up in the past, but also that she has no evidence to think that it will not tomorrow. She may not consciously apply the latter aspect of the method, but all cases of inductive inference that yield knowledge include not only positive instances but lack of evidence of negative ones. (It would be bad to infer that all my students will get As (even if true) on the basis that the first four I thought of will. I also must consider any evidence that other students may not get an A for the induction to be knowledge producing.) S knows that the sun will rise tomorrow if her method includes responsiveness to counterevidence, but not otherwise, in which case she may know that it is very likely that the sun will come up. With respect to the problem of higher-level knowledge, I trust that my response to Vogel above suffices to show that it is compatible with sensitivity.

I tentatively propose that knowledge is sensitive true belief formed by a reliable process. This theory entails the denial that knowledge is closed under known entailment. The counterintuitiveness of denying closure requires that I say something to explain how closure could fail, a task I leave to Chapter 6. In the meantime, we will examine another modal principle, safety, which says that a true belief that *p* is knowledge only if, in the nearby worlds where S believes that *p*, *p* is true. Safety is gaining momentum in the current literature, especially as a replacement for sensitivity in light of its alleged problems. In addition, it does not entail the denial of closure, and it may achieve even stronger anti-skeptical results. Safety deserves a fair and extended hearing.

4 Safety

I argued in Chapter 3 that sensitivity is not in fact incompatible with higher-level knowledge and inductive knowledge. As is fairly obvious, however, some philosophers disagree and count this as two strikes against sensitivity. The third strike – the strikeout pitch according to some – is that sensitivity theorists are committed to denying closure, a principle that many think is nonnegotiable. As I've said, I aim to take the sting out of non-closure in Chapter 6, but our present concern is a modal principle that, its proponents claim, avoids the problems inherent in a sensitivity-based account of knowledge. The principle is safety, and it says roughly that S knows that *p* only if, in nearby worlds where S believes that *p*, *p* is true. The principle admits of weaker and stronger readings, and we shall consider grounds for favoring each. Ultimately, however, I reject safety. First (§B), it is not entirely clear that there is a single version of safety that handles both veritic luck and inductive knowledge any better than sensitivity, but this is, in my view, simply a reason for *not favoring* safety over sensitivity (as opposed to a reason for rejecting safety outright). Second (§C), there are plausible counterexamples to the claim that safety is necessary for knowledge. Third (§D), as with sensitivity, to eliminate process luck, safety requires supplementation by some process reliabilist condition, but it just so happens that, whereas sensitivity is a natural complement to process reliabilism, safety conflicts with the fundamental intuition that one cannot know that one is not a BIV, and this is because of the general problem that it is in tension with the core rationale for process reliabilism, or so I shall argue.

A. Motivating safety

The rationale for safety is multifaceted. Sosa argues that sensitivity is incompatible with both higher-level knowledge and inductive knowledge.[1] We have assessed those arguments in Chapter 3 and found them not entirely convincing. Working under the assumption that sensitivity fails, however, Sosa replaces it with the principle of safety. This is his original description:

> Call a belief by S that p "safe" iff: S would believe that p only if it were so that p. (Alternatively, a belief by S is "safe" iff: S would not believe that p without it being the case that p; or, better, iff: as a matter of fact, though perhaps not as a matter of strict necessity, not easily would S believe that p without it being the case that p.)
>
> (Sosa 1999, 142)[2]

Duncan Pritchard defines safety thus: "A subject's belief in a contingent proposition ϕ is safe if, and only if, for a wide-range of near-by possible worlds, if the subject believes ϕ, then ϕ is true,"[3] and later, thus: "For all agents, ϕ, if an agent knows a contingent proposition, ϕ, then, in nearly all near-by possible worlds in which the agent forms her belief that ϕ in the same way as in the actual world, her belief is true."[4] The main difference between Pritchard's safety principles is the number or ratio of nearby worlds through which a subject's belief must be true (but not their distance from the actual world). This will be important below.

How does the appeal to safety solve the putative problems (for sensitivity) about inductive and higher-level knowledge? (I will sketch the answers here, but when we come to a critical assessment of safety in the next section, we will see that safety, suitably refined, is not obviously an improvement over sensitivity with respect to inductive knowledge after all.) Take Sosa's trash chute case. In most nearby worlds where S believes that the trash will soon arrive in the basement, it is true. In satisfying safety, S knows it will soon arrive in the basement. We get a similar result in the rookie case. In most nearby worlds where the policeman believes that the rookie failed to shoot the bullet down the barrel of the criminal's gun, it is true. With respect to higher-level knowledge, suppose that my belief that *p* is safe, from which it follows that it is true in most nearby worlds where I believe it. My belief that my belief that *p* is true, or not false, is also safe, because my first-level belief is safe, that is, true throughout nearby worlds, and in those worlds I believe that my first-level belief is true.[5]

That safety is purportedly more successful than sensitivity in handling these cases does not add up to a decisive reason for adopting it. Further motivation comes from the prospect of larger epistemological gains. One such gain, touted especially by Pritchard, is the elimination of veritic luck. Recall again the Mary problem. On the safety account, Mary does not know that her husband is in the living room, even though it is true, because "there are a number of near-by possible worlds where she forms her belief on the same basis and yet what she believes is no longer true" (Pritchard 2003, 118). By contrast, in bona fide cases of (safe) knowledge, "it cannot be a matter of veritic epistemic luck that her belief is true, since in nearly all near-by possible worlds where she forms her belief on the same basis she will form a true belief as a result" (Pritchard 2003, 118). It is not perfectly clear that this is so. We'll take up that issue below (§B).

An even larger epistemological gain is Sosa's and Pritchard's achievement of insight into Moore's response to the skeptic. Moore famously argued that, if he can show that at least two things exist, he can prove the existence of the external world. Holding up his hands, he showed that two things exist. Now, of course, someone who takes skepticism seriously might think this argument some kind of joke, but Moore claimed, first, that there is a distinction between knowing and proving, hence that, though he cannot prove his premises to the satisfaction of the skeptic, he can know them and use them as premises in a proof of the external world, and second, that one of the reasons it is plausible to say that he knows the premises is that his grounds for believing that he has hands are far stronger than any skeptical grounds for doubt.[6] Thus he understood quite well that a dyed-in-the-wool skeptic may not be convinced, but also that he was not *merely* thumbing his nose (or hand-waving) at the skeptic.

Notice, then, that Moore himself tacitly appeals to the closure principle against the skeptic. Moore knows that he has hands, knows that this entails, for instance, that he is not dreaming (or a BIV), and therefore he knows that he is not dreaming (or a BIV, and that there is an external world). So the neo-Moorean safety theorist kills three birds with one stone. First, unlike sensitivity, safety never takes into account very distant possible worlds. This is sometimes seen as a virtue in itself precisely because it is the very distant worlds that seem irrelevant to our knowledge statuses. In contrast, application of sensitivity for propositions like *I am not a BIV* requires consideration of the counterfactual: If it were not the case that I am not a BIV, I would not believe that I am not – consideration, that is, of a world where I am a BIV. Ignoring those worlds, safety easily accounts for our knowledge of ordinary propositions, which, if true, are true in nearby worlds – the only ones relevant to safety. Second, safety is compatible with closure, also because it does not take into account distant possible worlds. In the closest possible worlds where I believe I am not a BIV, it is true. This is cited as a *huge* advantage over sensitivity. Third, it follows from its ability to sustain closure that safety is compatible with knowledge that skeptical hypotheses are false, so the anti-skeptical result achieved by safety is even stronger than that accompanying sensitivity. (Although, as I said before, since it's difficult to see how one could obtain such knowledge, the safety theorist's ambition to maintain closure implies the denial of deeply held intuitive judgments about our lacking knowledge of the falsity of skeptical hypotheses.) One would be hard-pressed to deny the prima facie attractiveness of these features of safety.

B. Weak and strong safety; luck and induction

Is there any univocal version of safety that handles both the Mary problem (that is, veritic luck) and cases of inductive knowledge? John Greco offers a neat little dilemma that suggests otherwise.[7] The first step is to note the vagueness inherent in the safety principle; it can be construed very weakly, weakly, and strongly. Greco's dilemma involves only the weak and strong versions, but let us begin with the very weak version: "A subject's belief in a contingent proposition ϕ is safe if, and only if, for a wide-range of near-by possible worlds, if the subject believes ϕ, then ϕ is true" (Pritchard 2002a, "Moorean response," 295). This can be contrasted with both a slightly strengthened but nonetheless weak version and a strong version, where strength is determined by the ratio of nearby worlds through which one's belief must be true:

S's belief that (a contingent proposition) *p* is true is *weakly safe* if and only if, in nearly all nearby possible worlds in which S forms the belief that *p* in the same way as in the actual world, S's belief is true.

S's belief that (a contingent proposition) *p* is true is *strongly safe* if and only if, in all nearby possible worlds in which S forms the belief that *p* in the same way as in the actual world, S's belief is true.[8]

Suppose we favor very weak safety. Then the Mary problem constitutes an obvious counterexample, since surely there is a *wide range* of nearby possible worlds where Mary's vision-based belief is true; that is, Mary's belief satisfies very weak safety but she does not know. Very weak safety is simply too weak to eliminate obvious instances of luck. Greco claims that the same problem afflicts even weak safety. In *almost all* nearby worlds where Mary forms the belief that her husband is in the living room, it is true, because it is something of a fluke that her husband's brother is there and that she misidentifies him. In her original description of the case, Zagzebski says, "her belief has enough warrant to constitute knowledge when true and we can assume that it is almost always true."[9] "And yet," says Greco, "the belief is also veritically lucky, since there is a nearby world where her belief is formed in the same way and yet false" (Greco 2003a, "Virtue and luck," 360). An important issue in the Mary problem is whether there are more than just a few such possible worlds, in which case the weak safety principle is not satisfied, or extremely few, in which case it is. Greco, leaning heavily on Zagzebski's claim that Mary's belief "is almost always true," argues that weak safety does not go far enough in eliminating luck.

Well, clearly strong safety gets the right result in the Mary case. There is a nearby possible world where her belief, formed as it actually is, is false, and so it violates strong safety. But, Greco argues, if strong safety is required to eliminate luck, a problem about inductive knowledge arises. Take the trash chute case again. Surely there is at least one nearby world where the bag gets stuck, and in that world I believe it will soon be in the basement anyway. Notice that, though it is unlikely that the bag will get stuck, that does not necessarily mean that the possible world where it does is very distant. (Imagine that some prankster crams a mass of compressed foam in the chute which, when it expands, creates a clog. You know the stuff – it's fairly common; you know the guy, too – it's good that he rarely comes around here.) Compare the lottery case. Though it is extremely unlikely that any particular ticket will win, the possible world where it does is not distant, for it is just like the world where the ticket loses. Strong safety enjoys an advantage over sensitivity in accounting for inductive knowledge where the world in which the relevant proposition is false is not only unlikely but also very distant – where there is *no* nearby world where it is false. The trash chute case is not obviously like that, and so my belief that the bag will soon be in the basement may very well violate strong safety, from which it follows that I do not know that it will arrive in the basement even when it will.

Perhaps there is a bit of irony here, in that safety was originally introduced by Sosa partly as a response to this sort of problem with sensitivity. Or perhaps not. It may not be in principle impossible to find a middle ground, somewhere in the neighborhood of weak safety, or between weak and strong safety, one that accounts for luck in the Mary case but allows inductive knowledge. In addition, we have to be very careful to play by a consistent set of rules when evaluating the relative merits of sensitivity and safety. Recall that in the barn facsimile case, we said that if there are no fake barns within, say, hundreds of miles, the possible world where the thing Henry sees is a barn is not a close one. (Thus distance is not a metric of probability, but must take into account objective facts about the environment. In the lottery case, the probability that a given ticket will lose is very high, but that does not undermine the claim that the world where the ticket wins is a nearby one.) By parity of reasoning, we should say that if there are in fact no pranksters contemplating stuffing foam in this or any other chute in the neighborhood (city, state), then the possible world where there is such a prankster is not nearby. And if there is such a prankster, then even though strong safety is violated, it is plausible to say that I do not know that the bag will soon be in the basement anyway, in which case strong safety gets the right result.[10]

I will leave it as an open question whether Greco's dilemma for safety is ultimately seriously damaging, but at a minimum, Greco is right that it calls for further treatment by advocates of safety; he is right that there is a general tension in aiming to eliminate veritic luck while accounting for inductive knowledge:

> On the one hand, it seems that veritic luck is incompatible with knowledge, and that therefore we need some kind of condition on knowledge to eliminate the possibility of veritic luck. On the other hand, it seems that inductive knowledge is possible, and that inductive knowledge carries with it the possibility of veritic luck. … [This] is a perfectly general problem, faced by any plausible epistemology.
>
> (Greco 2003a, "Virtue and luck," 361)

Given the inconclusiveness of this issue, I present it not as a decisive reason for rejecting safety in particular, but as undermining safety's claim to superiority over sensitivity in handling both veritic luck and inductive knowledge. I believe there are better reasons for denying safety, but this story is a bit longer, to be told over the next three sections.

C. Is safety necessary for knowledge?

The discussion in the previous section shows that, to steer a course between veritic luck and the inability to account for inductive knowledge, safety must be construed fairly strongly. But this creates the possibility of counterexamples to the claim that safety is necessary for knowledge.

In a recent article, Juan Comesaña presents just such a case.

> There is a Halloween party at Andy's house, and I am invited. Andy's house is very difficult to find, so he hires Judy to stand at a crossroads and direct people towards the house. … Unbeknownst to me, Andy doesn't want Michael to go to the party, so he tells Judy that if she sees Michael she should tell him the same thing she tells everybody else (that the party is at the house down the left road), but she should immediately phone Andy so that the party can be moved to Adam's house, which is down the right road. I seriously consider disguising myself as Michael, but at the last moment I don't. When I get to the crossroads, I ask Judy where the party is, and she tells me that it is down the left road.[11]
>
> (HALLOWEEN PARTY)

On the one hand, it seems right to say that I know that the party is down the left road, whereas on the other my belief is unsafe – there are a number of nearby worlds where I believe that the party is down the left road but they are false worlds where I disguise myself as Michael. Perhaps a variant on the case will make it more plausible. Suppose that I was supposed to give Michael a ride to the party, and at the last minute I simply forgot. Had he been in the car, Judy would have told us just what she actually did, but the party would have been moved. My belief, based on the method of believing what Judy tells me, is unsafe because there are nearby worlds where it is false, but again, for all that, it seems wrong to say that I do not know. Sensitivity handles this case better, for in the nearest worlds where it is false that the party is down the left road, I do not believe it. The *nearest* such worlds are worlds where the party is somewhere else, or there is no party, and Judy tells me so. Worlds where Michael is in the car are more distant.[12]

One might think there is some wiggle room here for safety theorists. Perhaps a retreat to weak safety would help, since throughout a wide range of nearby worlds my belief that the party is down the left road is true, hence my belief constitutes knowledge. But in the previous section we saw how weak safety is too weak to eliminate genuine instances of veritic luck, as in the Mary problem. I submit, then, that the safety theorist faces a serious challenge in finding a reading of the principle that does not run aground on at least one of the difficulties presented in these last two sections. In the following two sections, I aim to compound the problems for the safety theorist.

D. Luck revisited: Safety requires a process reliability condition

Let me give the gist of the argument I will pursue in this section and the following one, an argument that informs the theory developed in the next chapter and which partly motivates my reasons for denying the closure principle. Pritchard (2003) argues that safety handles veritic luck better than process reliabilism or agent reliabilism, the latter being, as we will see, a version of virtue theory. For present purposes, let us focus on just process reliabilism with the understanding that Pritchard's claims apply to agent reliabilism as well. (This assumption gets a partial defense, below.) Pritchard argues further that this result undercuts the rationale for reliabilism, leaving the safety principle as the only requisite belief-truth link in its wake. I argue, in reply, that safety is susceptible to clear counterexamples involving process luck unless it, too, is supplemented by a process reliability condition, hence reliability of the belief-forming process is a necessary condition for knowledge.[13]

Based on this conclusion, in addition to the results of Chapter 3, *any* modal principle attaching solely to specific beliefs must be supplemented by process reliabilism. The question is: Should we choose sensitivity or safety to go with reliabilism? Safety theorists cite the compatibility of safety and closure as a virtue. But this implies that, on the safety view, one can know that one is not a BIV. Whereas sensitivity theorists are sometimes accused of epistemological revisionism in their denial of closure, I think the charge is in fact *more* forceful against safety theorists, because it is a founding intuitive judgment in epistemology that we do not know that radical skeptical hypotheses are false. Founding in what sense? Well, for one thing, the whole concern about the status of closure *derives* from the more basic judgments that we can know ordinary truths but not the denials of skeptical hypotheses. It should surely give us pause to be told that some theory squares anti-skepticism with closure, where the trick is turned by claiming we know that which we firmly believe we cannot know.

I further claim that, at the heart of reliabilism is the idea that knowledge requires responsiveness to the world, where what that comes to is an ability to distinguish truth from falsity. This point is not obvious given just the basic idea that reliability is simply truth-conduciveness, and so I shall say more below to develop it. If it is right, then it makes sense to think of sensitivity, which demands a capacity to distinguish truth from falsity, as a more natural complement to process reliabilism than safety, which makes no such demand.

We begin by looking at how the skeptic applies the closure principle in wielding the argument from ignorance:

(1) I do not know that I am not a BIV.
(2) If I do not know that I am not a BIV, then I do not know that I have hands.
(3) Therefore, I do not know that I have hands.

Notice that (2) is an expression of closure: If S knows that *p* (in this case, *I have hands*) and knows that *p* entails *q* (here, *I am not a BIV*), then S is at least in a strong enough position to know that *q*; if S does not know that *q* but knows that *p* entails *q*, then S does not know that *p*. (The argument from ignorance and its implications are treated in more detail in Chapter 6.) At this point, if we reject safety, our options are to deny closure or to accept widespread skepticism. I opt for the former. Knowledge is inconsistent with veritic and process luck; eliminating luck requires a process reliabilist condition; process reliabilism is part of a successful account of the nature of knowledge, whether knowledge exists or not; on a process reliabilist account, adequately refined, we (may) know many ordinary propositions but not the denials of skeptical hypotheses; therefore, closure fails. At the end of this chapter, and again in the next, I argue further that "adequately refined" motivates the addition of a sensitivity condition partly because sensitivity and process reliabilism are natural complements.

Let's take it from the top with Pritchard's challenge to reliabilism.[14] He claims that reliabilism may be part of an *explanation* for how we come to have knowledge, but, once we have a modal principle that eliminates veritic luck, which in his view is safety, process reliabilism no longer needs to be part of an *analysis* of knowledge:

> There is no inherent reason, however, why a safety-based account of knowledge should mention the reliability of processes at all. . . . [T]he appropriate moral to draw . . . does not seem to be that we need to keep supplementing the reliabilist thesis ad infinitum . . . but rather that we should simply accept that knowledge is, at root, just true belief that meets the safety principle.
>
> (Pritchard 2003, 119)

In this particular paper, Pritchard favors what we, following Greco, have called weak safety: "For all agents, ϕ, if an agent knows a contingent proposition, ϕ, then, in nearly all near-by possible worlds in which the agent forms her belief that ϕ in the same way as in the actual world, her belief is true" (Pritchard 2003, 118).

Pritchard argues that safety-based accounts of knowledge can handle cases of veritic epistemic luck where modalized process reliabilism and agent reliabilism fail. The superior ability of safety to eliminate veritic luck, it is claimed, undermines the motivation to supplement an analysis of knowledge by agent, process, or faculty reliabilism, or by anything else, such as an appeal to epistemic virtue.[15] This is so because process reliabilism and virtue theory are themselves partly motivated by the need to eliminate veritic luck. Pritchard's point, then, is that if a safety-based account can manage better on this score by itself, reliabilists and virtue theorists must offer a further rationale for claiming that reliable belief-forming processes or beliefs deriving from intellectual virtues, respectively, are necessary for knowledge. (I should note that, in the paper at issue, Pritchard's focus is on virtue theory. Still, the main claim under discussion here is explicitly directed at both process reliabilism and virtue theory, as the first indented quotation in the previous paragraph shows.) Pritchard may well be right that his favored notion of safety, which does not explicitly mention reliability or virtues, fares better than its reliabilist competitors in eliminating veritic luck, but, to eliminate process luck, too, it must be supplemented by process reliabilism or by something very like it.

Let us retrace the dialectic leading to Pritchard's view. We saw in Chapter 2 that actual world process reliabilism allows for far too much knowledge-precluding epistemic luck. Consider Greco's benevolent demon:[16]

> René thinks he can beat the roulette table with a system he has devised. Reasoning according to the Gambler's Fallacy, he believes that numbers which have not come up for long strings are likely to come up next. However, unlike Descartes' demon victim, our René has a demon helper: every time René forms a belief that a number will come up next, the demon arranges reality so as to make the belief come out true.

This process is truth-conducive in René's world, but it is false that René knows which numbers will come up next. René's beliefs are luckily true. Plantinga's brain lesion case similarly demonstrates the deficiency of actual world process reliabilism.[17]

I argued in Chapter 2 that a truly reliable process is one that produces true beliefs not only in the actual world but also throughout nearby possible worlds. ("Throughout" leaves open the question of how *many* nearby worlds, that is, the ratio of nearby worlds where the belief is true, and so calls for refinement. I leave that until Chapter 5.) It is plausible to say that René's demon does not exist (or at least does not help) in nearby worlds, and that in most nearby possible worlds, brain lesions do not cause true brain lesion beliefs. This basic idea is not proprietary to process reliabilism. Agent reliabilism and virtue reliabilism hold that a reliable process is one grounded in stable cognitive (or intellectual) dispositions of the agent.[18] All of these views account for the lack of knowledge in the counterexamples to actual world process reliabilism by noting that the putatively reliable processes in these examples are not stable in the relevant respects – they do not produce true beliefs in most nearby possible worlds.

Though agent and virtue reliabilisms make only implicit reference to process modalization – the *central* idea being stable dispositions – I shall lump them in with modalized process reliabilism because, first, they achieve similar results in their amendments to actual world process reliabilism, and second, they are susceptible to similar counterexamples. On this second point, we return, as you might have guessed, to the Mary problem (Zagzebski 1996, 285ff.). Mary's belief is true and formed by a reliable process, one that produces true beliefs not only in the actual world, but throughout the range of nearby possible worlds. (Similarly, her vision-based belief is formed by a stable disposition and a reliable character trait, in Sosa's sense, where "character trait" does not necessarily imply a trait for which an agent is responsible.) But her belief is veritically lucky.

This is where Pritchard invokes safety. Mary's belief is not safe, and therefore does not constitute knowledge, because she would believe that her husband is in the living room in nearby worlds where it is false. (Let us set aside the potential problems for safety revealed by Greco and through the Halloween Party case, discussed above, and assume that the applicable version of safety is strong enough to deny that Mary's lucky belief is knowledge while not too strong to make inductive knowledge impossible. The issue I want to raise has nothing to do with the strength of the safety principle.) Now we are in a position to take up Pritchard's challenge: "There is no inherent reason, however, why a safety-based account of knowledge should mention the reliability of processes at all, whether those at issue in agent reliabilism or those at issue in the modalized process reliabilist story" (Pritchard 2003, 119). I agree with Pritchard that modalizing belief-forming processes does not suffice, given the Mary counterexample. I also agree that modalization thus must apply to individual beliefs, as safety would have it. However, I think that counterexamples can be constructed in such a way as to motivate process reliabilism within this framework.

It will take a couple of iterations to build counterexamples that really "stick" to the safety-based account, for many prima facie counterexamples can be handled by indexing safety to one's method of forming belief. Starting with Sosa's characterization of safety, which lacks mention of methods, helps us to see why we need that amendment:

> Call a belief by S that p "safe" iff: S would believe that p only if it were so that p. (Alternatively, a belief by S that p is "safe" iff: S would not believe that p without it being the case that p; or, better, iff: as a matter of fact, though perhaps not as a matter of strict necessity, not easily would S believe that p without it being the case that p.)
>
> (Sosa 1999, 142)

A belief is safe if and only if, throughout most nearby worlds, if S believes that *p*, *p* is true. (Let us assume that) Mary's belief violates safety, since there are nearby worlds where she believes that her husband is in the living room but he is not, and this explains why Mary does not know.

Pritchard ("Moorean response," note 18) cites Sosa's construal of safety approvingly: "I take the formulation offered here to capture the spirit of all three" of Sosa's formulations, above. But Sosa's rather austere statements of safety still yield counterexamples. Before we come to the counterexamples that stick, it will be useful to see how safety needs to be amended by indexing its application to the method of forming belief, as was the case with sensitivity. Consider John. John is in your Introduction to Philosophy course, in which you're discussing Descartes' dreaming argument. Aiming to be "relevant" to students' interests, you explain how, just as Descartes' argument threatens ordinary empirical knowledge, so too does the worry that one may be in the Matrix. (In the movie *The Matrix*, the Matrix is a computer-generated virtual reality that conforms closely to the world as we think it is, though in fact human beings are essentially, to make a long story short, BIVs.) John immediately forms the true belief (q) that he is not in the Matrix. The belief also appears to be safe – in (many, most, and all) nearby worlds where John believes that q, q is true. Matrix worlds are, we may suppose, very distant.

But I've said nothing about John's method or way of believing that q. As it happens, John just does not like Keanu Reeves (KEANU REEVES). His girlfriend cannot stop talking about Keanu's good looks, and John thinks that Reeves cannot act. John believes that he is not in the Matrix because he associates the very idea with an actor that he hates. (We might suppose, too, that he believes he is not dreaming because he associates the very idea with a class he hates. We've all been *there*; we all know John.)

Does John *know* that q? It seems utterly implausible that he does because, although his belief is (alas) entirely psychologically plausible and apparently safe, it is not formed by a reliable process. This kind of case suggests that safety needs to be supplemented by process reliabilism – the result I am after – but there is another potential move for the safety theorist to make in reply, namely, that John's belief is *not safe after all* because the safety principle needs to be indexed to the specific method of belief formation. In the following chapter, I offer a criterion for distinguishing belief-forming process types from the narrow methods by which particular beliefs are formed. For present purposes, what is salient is that the former must be specific-belief-content-neutral whereas methods are content-specific (as we saw in Chapter 3). Thus John's method may be something like: *forming a belief about whether I am in the Matrix based on my emotional reaction to the star in the movie.* (John need not know that this is what he's doing.) This particular method might plausibly be said not to produce a safe belief, since there are nearby worlds where there is no such movie or in which Keanu Reeves does not appear in it, hence where John forms the belief, based on his actual method, that he *is* in the Matrix. (We may suppose that John is paranoid.[19]) There may be some other method by which John would believe that he is not in the Matrix, but relative to this method, his belief is not safe, and that is why he does not know.[20]

We may thus conclude that John's belief is unsafe only if the safety principle is indexed to the particular method of forming belief. Another characterization of safety makes this explicit:

> Safety$_{\text{Pritchard}}$:
> For all agents, ϕ, if an agent knows a contingent proposition, ϕ, then, in nearly all near-by possible worlds in which the agent forms her belief that ϕ *in the same way* as in the actual world, her belief is true.
> (Pritchard 2003, 118, my emphasis)

But this still does not suffice, for note that John's belief satisfies even this condition. In all the nearby worlds where he believes that he is not in the Matrix, based on his actual method, it is true. We said that John's belief may be deemed unsafe because, though he actually did not, John might just as easily have formed the false belief that he *is* in the Matrix, or failed to believe that he is not, because there are nearby worlds where there is no such movie or Keanu Reeves is not in it. If this move is to redeem safety, then Pritchard's characterization of the principle needs a little tweaking, thus:

> Revised Safety$_{\text{Pritchard}}$:
> For all agents, ϕ, if an agent knows a contingent proposition, ϕ, then, in nearly all near-by possible worlds in which the agent forms her belief ABOUT ϕ in the same way as in the actual world, her belief is true.

According to safety$_{\text{Pritchard}}$, John's belief *that* he is not in the Matrix could still be deemed safe, since in nearly all nearby worlds in which he forms *that* belief, it is true. However, his doxastic attitude toward, or *about* the proposition that he is not in the Matrix may not be safe, since he might have adopted a different attitude toward that proposition in some nearby world where he bases his belief on the method – *forming a belief about whether I am in the Matrix based on my emotional reaction to the star in the movie.* Construing safety this way brings it closer to Pritchard's characterization of *super-safety*:

> An agent has a super-safe belief in the contingent proposition ϕ if, and only if, that agent has a true belief in ϕ and her belief in ϕ 'tracks' the truth as to whether ϕ across a wide range of near-by possible worlds.
> (Pritchard 2002a, "Moorean response," 297)[21]

Pritchard seems to prefer super-safety to safety: "What this formulation adds to Sosa's notion of safety is the contention that knowledge does not just require an agent's belief to track the truth across the realm of safety, but also that the agent's *non*-belief should do likewise" (Pritchard 2002a, "Moorean response," 297). Super-safety adds to safety something like a sensitivity requirement, in that it requires that one not hold the belief in nearby worlds where it is false. But this is not the Nozickean sensitivity subjunctive conditional that I have tentatively accepted as a constituent condition for knowledge, because, unlike Nozick's view, only worlds close to the actual world are relevant to determining super-safety, even with respect to beliefs that skeptical hypotheses are false. This makes possible the additional anti-skeptical results: Since I believe I am not a BIV throughout close possible worlds – there are none where I do not believe it – and it is true (let us assume) throughout those worlds, I know it. (See Pritchard 2002a, "Moorean response," §V.) My amendment to $\text{safety}_{\text{Pritchard}}$ is meant to capture this element of super-safety – one's attitude toward a proposition, not just one's belief that it is *true*, must "track" the truth.

For the sake of explicitness, we want super-safety to be relativized to the method of belief formation, as demanded in the case of John. The remaining question is: Is relativization to way of believing enough to save safety from counterexamples?

Suppose that Smith believes that the earth revolves around the sun. As a sun worshipper, she believes this solely for religious reasons (SUN WORSHIPPER). Presumably, worlds where this is false are more distant than ones in which Smith even exists. Thus her belief is safe – in all the nearby worlds where she believes it, it's true. Even when we include ways of believing in our safety principle and individuate ways (or methods) of believing narrowly, in particular, content-specifically, as we did for sensitivity, then again Smith's belief is safe. Her way of believing that the earth revolves around the sun is based on her heliocentric religious beliefs, and there are no nearby worlds where her particular belief *about* the proposition that the earth revolves around the sun, based on religious heliocentrism, is false. And surely it is not unfair to Pritchard to individuate ways and methods content-specifically, for were we to broaden the method type, it would threaten a lapse into process (or agent) reliabilism, which is precisely what Pritchard claims is unnecessary in the analysis of knowledge. There *is* a way to block these counterexamples, but only by appealing to belief-forming *processes* that produce safe beliefs (or, if you like, by appealing to stable cognitive characteristics that have the same modal status). One could then claim that it is the general unreliability of forming beliefs based on heliocentric religious views that precludes knowledge in the Smith case – that Smith's beliefs involve process luck. Smith's belief, taken independently of her way of believing, is safe. Smith's belief, assessed relative to her way of believing, is also safe, but Smith's religious views do not constitute a reliable belief-forming process, and *that* is why she does not know. That is, Smith forms many false beliefs based on her religious views, for instance, that droughts are explained by reference to the sun god's anger, and that the sun is the center of the universe. I tentatively conclude that safety is capable of eliminating process luck only in combination with some condition akin to modalized process reliabilism.

The conclusion is only tentative because there is yet one more caveat to be entered on behalf of Pritchard's safety-based account, namely, that the safety principle applies only to "deeply contingent" propositions, those that are not necessary in any sense, logical, physical, or nomological.[22] Though the proposition that the earth revolves around the sun is not obviously necessary even in any of these ways, it is, as I have described it, a necessary condition of Smith having such a belief insofar as worlds where it is false are more distant than ones where Smith exists, a fortiori they are more distant than ones where she has the relevant belief. Because the safety principle is not intended to apply to such a proposition, it is inappropriate to count Smith's belief as a case of knowledge according to safety.

There is nothing conceptually untoward about this move, but one question that it raises is whether it undermines the greater anti-skeptical results offered by safety theorists. If it is in fact true that I am not a BIV, it may not be a deeply contingent fact in the required sense, in which case safety would not apply, and I would not know that the proposition is true. Suppose, for example, that Davidson's argument that there are no psycho-physical laws is cogent, or at any rate that its conclusion is true.[23] Then it would be impossible for anyone who knew all the laws of nature to activate my brain states to produce the desired beliefs, and so I *could not* be a BIV in the relevant sense – a BIV "programmed" to have ordinary world beliefs. Perhaps an omniscient being could produce in me, directly, whatever beliefs it desired me to have, but that is to imagine another possible world that is extremely remote (at least as remote, intuitively, as a world where the earth does not revolve around the sun), in which case the proposition that I am not a BIV whose beliefs are manipulated by an omniscient being would seem anything but deeply contingent. The safety principle would then be *inapplicable*, and I would not be able to know that proposition.

The part of this argument against the safety-based account of knowledge of anti-skeptical hypotheses which appeals to Davidson's anomalism is somewhat contentious, but the basic argument illustrates a more general point, which is that the more bizarre the skeptical scenarios, the closer they come to being necessarily false (in some sense of "necessarily" – at least as opposed to deeply contingent). Appeal to safety as a principle applying only to deeply contingent propositions is thus powerless to account for our knowledge that such skeptical hypotheses are false. This is true even where the method by which one comes to believe the denials of skeptical hypotheses is through inference from the ordinary empirical propositions that one believes – ones that are known to be incompatible with skeptical scenarios.

The conclusion I want to draw is that neither safety nor sensitivity provides a basis for our understanding of knowledge of necessary truths, which is one area where an appeal to reliable belief-forming processes enters the picture. Relatedly, it is not entirely clear that safety is compatible with knowledge that radical skeptical hypotheses are false, and if that's right, it undercuts one of the primary advantages claimed for safety over sensitivity.

(I need not insist on this second point, though I think it's quite significant. My main objective is to secure the first point, that is, to motivate a process reliabilist feature in our theory of knowledge. On the other hand, if safety *is* compatible with knowledge that radical skeptical hypotheses are false, it undermines the basic intuitive judgment that we lack such knowledge.)

Even apart from necessary propositions, there are examples that militate in favor of process reliabilism. The Haveit case is one of them. Haveit wants me to believe truly that someone in the office owns a Ford, so he provides me with false evidence that Nogot has one, when in fact Haveit owns a Ford. In the nearby worlds where I believe that someone in the office owns a Ford, based on the method of *believing that someone in the office owns a Ford on the basis of evidence provided by Haveit*, it is true. Haveit, by stipulation, would only give me false grounds if it were true that someone in the office owns a Ford. I do not know that someone owns a Ford, but my belief, based on the specific method actually used, is safe. This example of process luck created difficulty for sensitivity, too, and the proposed solution involved an appeal to the general unreliability of Haveit. (How exactly to explicate the unreliability is a topic for the next chapter.)

To this point, we have no decisive reason for rejecting safety in favor of sensitivity, though with the additional problems presented in §§B and C, the evidence against safety is mounting. We have seen that both modal principles leave open the possibility of process luck and thus require supplementation by process reliabilism (or an agent reliabilist or virtue theoretic counterpart, which for the rest of this chapter I shall take to be understood). We have seen that a plausible safety-based reply to the problematic case of Smith believing that the earth revolves around the sun that avoids an explanation in terms of process reliabilism – the reply that safety applies only to deeply contingent propositions – creates a tension with the safety-based account of knowledge of the denials of skeptical hypotheses because many such propositions are not obviously deeply contingent. What further considerations might favor sensitivity?

E. Is reliability compatible with knowledge of the denials of skeptical hypotheses?

Eliminating egregious cases of process luck entails a process reliabilism condition, whether one accepts sensitivity or safety. Safety theorists claim the advantage of upholding closure over those who favor sensitivity, and with it knowledge of the denials of skeptical hypotheses. I have just argued that safety may not be as potent as originally advertised in this regard. Here I shall switch tactics a bit, and argue that the safety theorist's claim to accommodate knowledge of the denials of radical skeptical hypotheses is seriously problematic, both intuitively and because it allows for knowledge that is not responsive to the world, which is in tension with a core notion of reliabilism. If this is right, we should couple the requisite process reliabilism feature of our theory with sensitivity rather than safety.

First off, I had better enter into the proceedings my awareness that reliability means different things to different people. Appealing to any favored notion is bound to seem question-begging. But we can make some headway toward a generally shared concept of reliability by noting what it cannot be in the present context. In the introduction, we distinguished global, process reliabilism from local reliabilism. Safety and sensitivity are versions of the latter. It was the work of the previous section (and of part of the previous chapter) to show that local reliability is insufficient for knowledge. So it would be wrongheaded to say, at this point, "by 'reliability', I mean satisfying the safety principle" (or "by 'reliability', I mean satisfying the sensitivity principle"). No. By "reliability" we all mean process reliabilism.

1. Reliability, responsiveness, and discrimination

In her recent book, Sherrilyn Roush presents a "direction of fit" argument against safety. The concern is that "knowledge is a matter of responsiveness to the way the world is and safety makes a demand in the opposite direction" (Roush 2005, 122). The sun worshipper example shows this. Smith has a safe belief that accords with, but is not in any way responsive to, the facts. Sensitivity, however, is a kind of responsiveness to the world, for it requires that, if the world were different than it is, one would not have the belief that one does. I agree with Roush that responsiveness to the world is an important part of knowledge, and it is clear that safety does not require responsiveness. This last point helps us to understand (or diagnose) how the safety theorist's claim that we can know that radical skeptical hypotheses are false contradicts our intuitive judgments about such cases. On the safety account, I can know that I am not a BIV because the realm of safety does not extend out to BIV worlds, hence my belief that I am not a BIV allegedly constitutes knowledge because it is true throughout nearby worlds. But this is just to admit that my belief that I am not a BIV is not responsive to the possibility that I am, which would seem to constitute the crucial question on this issue.

I would be content to let this point, together with the arguments from §§B through D, comprise my case against safety. But I've also claimed that sensitivity and process reliabilism more naturally complement each other than do safety and process reliabilism, and this demands an explanation, one that centers on the notion of responsiveness, or, at any rate, its close cousin – the capacity to distinguish.

Alvin Goldman, who might be considered the guru of reliabilism, originally described reliability in these terms: "To be reliable, a cognitive mechanism must enable a person to discriminate or differentiate between incompatible states of affairs. It must operate in such a way that incompatible states of the world would generate different cognitive responses."[24] Following Goldman, McGinn says:

> [T]he underlying [basic] notion [of knowledge] is that of what might be called *distinguishing* knowledge, i.e. knowing one thing . . . *from* another. The result is a unified theory of knowledge in which the notion of *discrimination* is central and basic. This kind of theory is by no means novel – it is, indeed, a variant of what have come to be called *reliability* theories of knowledge.[25]

I want to highlight two points. First, McGinn claims that discrimination is central and basic, from which it follows that any proposed cognitive process that cannot discriminate truth from false counter-possibilities in its proprietary domain is not knowledge producing. Second, McGinn and Goldman both claim that discrimination *just is* reliability.

On the one hand, I do not see why anyone would want to dispute that discrimination or differentiation is a constitutive feature of reliability, while on the other hand I cannot imagine what any of this has to do with safety, at least as that principle applies to our beliefs that radical skeptical scenarios do not obtain. (Again, I am setting aside, for present purposes, the problem with the *applicability* of safety to beliefs that skeptical hypotheses are false. We may suppose here, then, that the falsity of skeptical hypotheses is relevantly contingent such that safety applies, which makes possible the safety-based account of such knowledge and thus the defense of closure.) It is agreed on all sides, based simply on *ex hypothesi* stipulation, that if S were a BIV, she would not believe she is. She has no cognitive mechanism that discriminates between the incompatible states of affairs in question – between what Hill calls the real-world hypothesis and skeptical hypotheses.[26] If we take a discriminative capacity to be the mark of reliability, it follows that beliefs in the denials of skeptical hypothesis are not reliably formed.[27] This is consistent with sensitivity – if it were false that S is not a BIV, she would believe it anyway.

The obvious problem with construing reliability in discriminative terms is that, officially, reliability is nothing other than truth-conduciveness – a process is reliable if it produces mostly true beliefs. How is discrimination implied? Well, recall from the benevolent demon case that actual world reliabilism is insufficient to capture the sense in which a process is truly reliable. The demon arranges reality to fit my wish-based beliefs, which is an indication that my beliefs are not responsive to the world. When we modalize the concept of a reliable process, requiring truth-conduciveness throughout nearby worlds, it ensures at least some level of responsiveness – some capacity to distinguish truth from falsity. Wishful thinking is not truth-conducive in any nearby world where my demon does not exist, or where it does not help me. A modally reliable process, on the other hand, produces true beliefs independently of any possible quirky features of the actual world. Because it requires truth-conduciveness not only in the actual world, it demands that the belief outputs of a process discriminate between the actual world and ways the world might have been; it requires responsiveness to the world. Here's another way to put the point. Though truth-conduciveness forms the basis of reliabilism, a process is genuinely reliable only if it eliminates process luck, and any process capable of eliminating luck must be responsive to the world, able to distinguish truth from falsity.

The central conclusion I draw from the considerations presented in this section is that sensitivity is better suited to the reliabilist program than is safety, since the latter does not demand responsiveness to the world, or discrimination. It is this lacuna in the safety account that provides a diagnosis of the concern that safety undermines the intuitive judgment that we do not know that skeptical hypotheses are false. How could I know that I am not a BIV when I have absolutely no way of telling the difference between the world as I believe it to be and a BIV world?

However, we must be cautious here. One obvious answer to this question is, "through deduction from known premises." Since deduction constitutes a reliable process, why shouldn't the reliability theorist also claim that we can know the denials of skeptical hypotheses? I know that *p* (for some quotidian *p*), that *p* entails that I am not a BIV, and therefore that I am not a BIV. Given the overwhelmingly plausible claim that I do not know that I am not a BIV, there must be some explanation of how deduction cannot deliver knowledge in this case – of how closure fails. In Chapter 6, I argue, based partly on considerations presented here, which are meant to show that reliability requires responsiveness to the world, that questions about our knowledge statuses raised by skeptical hypotheses introduce a kind of loophole in the reliabilist account, which at its core demands responsiveness, allowing the possibility of *unresponsive* knowledge that skeptical hypotheses are false. The addition of a sensitivity condition closes the loophole, restoring our intuitive judgment that we do not know that skeptical hypotheses are false. I argue further that closure fails because the warrant sufficient for knowledge of ordinary propositions does not suffice for knowledge that skeptical hypotheses are false, where an externalist account of warrant is crucial for understanding this point.

2. *On contrastivism*

I am obviously putting a lot of weight on discrimination as a mark of knowledge. Before wrapping up this chapter, then, I would like to digress briefly to explain why I favor a modalized epistemology rather than another up-and-coming theory that also appeals to discrimination but avoids the complications raised by modalism. The theory is contrastivism and is forcefully defended by Jonathan Schaffer.[28] "The contrastivist says that knowledge is a *ternary, contrastive* relation: *s* knows that *p* rather than *q*," as opposed to the more standard binary relation, *s* knows that *p*.[29] For example, since I can discriminate zebras from all the other animals in the zoo, it is true to say that I know that it is a zebra rather than some other kind of zoo animal, but false to say that I know it is a zebra rather than a cleverly disguised mule, and so false to say, more simply, that I know it is a zebra.

Schaffer claims that the contrastive account of knowledge requires five stages of development: "One needs to report the *use* of knowledge ascriptions (§1), limn the *structure* of the knowledge relation (§2), show how the ascriptions *express* the relation (§3), analyze or otherwise *illuminate* the relation (§4), and resolve outstanding *paradoxes* (§5)" (Schaffer 2005, 235; emphases and section numbers are Schaffer's).

According to Schaffer, the use of a knowledge ascription is to certify that the subject is able to answer the question guiding the inquiry, hence it also serves to keep score or gauge the progress of the inquiry. (Schaffer offers arguments to support contrastivism throughout his development of it, but I shall stick for the most part to the essentials of the view itself.) Second, the "knowledge relation has the ternary, contrastive structure: K*spq*" (239). Suppose that the question, explicitly asked or implicit in an inquiry, is: "Is there a *goldfinch* in the garden?" The proper structure for a knowledge proposition might be *There is a goldfinch in the garden, rather than a raven.* Suppose instead that the question is: "Is there a goldfinch in the *garden*, or at the *neighbor's*?" Then the right structure might be *There is a goldfinch in the garden, rather than at the neighbor's* (241). In both cases, what is actually uttered might be, simply, "There is a goldfinch in the garden." The third stage of Schaffer's account is "to show how knowledge ascriptions *express* the knowledge relation" (244), and his answer is that "[k]nowledge ascriptions encode K*spq*, by encoding relations to questions" (244). Having argued that questions present contrasts, in this section of his paper Shaffer explains the mechanisms by which knowledge ascriptions encode question-relativity. For example, "if one says, 'I know who stole the bicycle', then the embedded question 'who stole the bicycle?' presents a set of alternatives, such as: {Mary stole the bicycle, Peter stole the bicycle, Paul stole the bicycle}" (246). To know that *p*, say, *Mary stole the bicycle*, is to know that *p* rather than *q* – rather than *Peter or Paul stole the bicycle.* Fourth, Schaffer offers the following analysis of knowledge: "K*spq* iff: (i) *p*, (ii) *s* has proof that *p* rather than *q*, and (iii) *s* is certain that *p* rather than *q*, on the basis of (ii)" (255).[30] Fifth and finally, contrastivism, through its puzzle-solving assignment of truth-conditions, resolves the skeptical paradox, which hinges on closure. Though we will look more closely at the reliabilist explanation of closure *failure* in Chapter 6, it is important to see how the contrastivist both upholds closure and achieves some anti-skeptical results, for if the account works, its ability to maintain the intuitively correct closure principle is a nice feature.

Let us look again, then, at a version of the skeptic's argument from ignorance, meant to undermine knowledge of ordinary propositions:

(1) I do not know that I am not a BIV.
(2) If I do not know that I am not a BIV, then I do not know that I have hands.
(3) Therefore, I do not know that I have hands.

The second premise is grounded in an application of the principle of closure:

(C1) I know that I have hands.
(C2) I know that if I have hands, then I am not a BIV.
(C3) Therefore, I know that I am not a BIV.

It follows that if I know that I have hands, then I know that I am not a BIV, and by contraposition, if I do not know that I am not a BIV, then I do not know that I have hands – (2). Either (a) the closure principle is false; (b) we accept that we do not know many, if any, ordinary propositions; (c) we accept that we know that skeptical hypotheses are false, which is not an option for discrimination based accounts; or (d) we face the following paradox:

(P1) I know that I have hands.
(P2) If I do not know that I am not a BIV, then I do not know that I have hands.
(P3) I do not know that I am not a BIV (which, with P2, entails that I do *not* know that I have hands).

The contrastivist, rejecting the simple binary structure of knowledge propositions, resolves the paradox by claiming that the proper structure of propositions like (C1) is something like *I know that I have hands rather than being an amputee.* This bit of discriminatory knowledge does not license the second premise of the closure principle argument. It is false that *I know that if I have hands rather than being an amputee then I have hands rather than being a BIV,* and so there is no relevant second premise to ground the conclusion that *I know that I have hands rather than being a BIV,* hence that *I know that I am not a BIV.* We need not deny that the closure argument is (informally) valid; rather, it is simply unsound.

Thus on the contrastive account, the ternary structure of the relevant propositions resolves the skeptical paradox, undermines the skeptical consequence that we know nothing, and (as just shown) maintains closure. (P1*) I know that I have hands *rather than* being an amputee. (P2*) If I do not know that I am not a BIV *rather than* a BIV, then I do not know that I am handed *rather than* a BIV. (P3*) I do not know that I am handed *rather than* a BIV. Each of these reinterpretations of the original propositions from the skeptical paradox is true, the account makes room for ordinary empirical knowledge, and the paradox vanishes. (See Shaffer 2005, 263–4.)

Much of the contrastive account of knowledge hinges on the idea that it "serves our goal of *scoring inquiry*" (237). In the context of inquiry there is always some explicit or implicit question to which knowledge ascriptions and utterances are responses. Suppose again that the question is "Who stole the bicycle?" It is assumed that a bicycle is stolen, and the reply that Mary stole it means that Mary, rather than someone else, stole it. The binary proposition, *Mary stole the bicycle*, conceals this, and thus the binary structure "seems too impoverished to provide a stable account of partial success in answering, inquiry, and discrimination" (244). Contrastivism is thus driven by inquiry and its progress, the suggestion seeming to be that if there is no explicit or implicit inquiry, knowledge ascriptions are at best inert, at worst meaningless.

Should a fully general account of knowledge be tethered to the inquiry-and-ascription model? Does it capture the notion of *knowledge* itself (as opposed to its ascription) when there is no guiding question? If someone were to ask me whether I know what that (demonstrated) thing is in my front yard, I might say that I know it is a large tree, and in this context it might seem right to say that I know it is a large tree rather than a bush. But outside of any implicit or explicit inquiry, when I look out the window and see the large tree, I know that it is a large tree rather than anything else, and that it is in the front yard rather than anywhere else. It seems that much of what I know is like that. I know, for instance, that my computer weighs more than my paper notebook, that my guitar has six strings, that my mother's name is Dorothy, and that two and two is four. I can think these things, even outside a context of inquiry, which may at times be somewhat pointless, though not meaningless, and certainly not untrue, and in thinking or saying them, even without a specific, explicit or implicit guiding question, I am thinking or saying something true independently of any implied contrast. It is not merely that I know that my computer weighs more than my notebook rather than weighing more than something else, or that my guitar has six strings rather than some other number or rather than my friend's guitar having six strings, or that my mother's name is Dorothy rather than Debra, or that two and two is four rather than five, or finally, that there is a large tree in the front yard, rather than a small bush being there or rather than in the back yard. All of these contrastive formulations are indeed true, but I also know that these propositions are true and meaningful in their plain vanilla binary form.[31]

The contrastivist might reply that this is all beside the point, for there is in fact an implicit inquiry when I self ascribe knowledge that there is a large tree in the front yard ("What's that?"), and inquiry always implies a set of relevant contrasts. This is a contentious point, but let us play along for the moment. So rather than trying to circumvent the contrastivist ternary propositional structure by appealing to putative cases where there is no implied question, which for the sake of argument let us agree do not exist, we'll subscribe to the inquiry model and thus face a head-on confrontation with the contrastivist proposal.

The contrastivist claims that the relevant alternatives one must be able to rule out "are generated by an explicit or implicit question" (267). I shall argue that this is implausible, and that, without some *other* way of understanding what makes an alternative relevant, the inquiry-based model fails to achieve the desired combination of anti-skepticism and closure maintenance. Let's first see why the contrastivist account of relevant alternatives is wrong. Suppose that the question is: "Goldfinch or raven?" (244; GOLDFINCH). The way the question is focused generates the (two) relevant alternatives. Seeing that the bird is yellow, I conclude that it is a goldfinch. The contrastivist proposal implies that the inability to answer the question, "Goldfinch or canary?" does not undermine the ability to answer, hence knowledge of the answer to, the question, "Goldfinch or raven?" This may be a correct account of the answers I am in a position to provide, but even if so, it is not much of an account of knowledge. If I "know" only that it is a goldfinch rather than a raven on the assumption that it's one or the other, then I do not know, in *any* plausible sense of that term, that it is a goldfinch, or even, it seems to me, that it is a goldfinch rather than a raven. I indeed know that it is not a raven, but not that it is a goldfinch rather than a raven. It is not enough to say that the possibility that it is a canary is irrelevant because it was not generated by the inquiry.

Presumably, it is because one already *knows* that the bird is not a canary that the inquiry focuses on it being a goldfinch or raven. If so, it is not that the possibility that it is a canary is irrelevant. Rather, it is relevant but ruled out. If, on the other hand, "canary?" has not been at least implicitly ruled out (which it would be if, for example, canaries don't live around here and I know it), it may still represent a relevant alternative, regardless of how the question guiding the inquiry is posed. In this case, it is utterly misleading to say that I know that it is a goldfinch rather than a raven. Either I know that it is a goldfinch and no other kind of bird, or I don't know what specific kind of bird it is at all. If we allow relevance to be determined by the nature of the query in a given context, it follows that my "contrastive knowledge" – that I can exclude "raven?" – is compatible with not really knowing much at all about the kind of bird because there may be relevant alternatives that the question fails to acknowledge. This strongly suggests that we need a better story about relevant alternatives.[32]

So far we see that the alternatives generated by an inquiry-guiding question do not exhaust all the relevant ones. Unexcluded relevant alternatives can undermine even contrastive knowledge claims. Now suppose the question is: Do I have hands, or am I an amputee? The alternatives generated by the question are clear, but do they exhaust all the relevant ones? Suppose that, while asleep, my arms were tied behind my back and my hands numbed by a local anesthetic, so that I can now neither see nor feel them. Parallel to the goldfinch case, while I might have evidence that I am not an amputee (or that it is not a raven) – my mother tells me I have never had a limb cut off (or I see that the bird is yellow) – this may not be enough for even contrastive knowledge. Unless I can rule out other relevant alternatives, for instance that I was born without hands or they were blown off in an explosion, it is utterly misleading to say that I know that I have hands rather than being an amputee, for I may not know that I have hands at all.

Now, if the only contrastivist proposal for determining relevant alternatives is that they are generated by focused questions, and if I am right that such questions fail to account for all the relevant alternatives, then the contrastivist position on relevant alternatives is wrong. Worse still, this makes the contrastivist vulnerable to the following criticism: "Because focused questions fail to raise all the relevant alternatives, you need some other grounds for denying that far-off skeptical possibilities are relevant (because you need some other criterion for relevance)." Neither the possibility that I was born without hands nor the possibility that I am a BIV are relevant alternatives generated by the question: Do I have hands, or am I an amputee? So what is the difference between them? The contrastivist cannot say. And now the worry is not merely that the contrastivist implausibly attributes knowledge in cases like the goldfinch (because some relevant alternatives are ignored in the context of inquiry), but also that, in light of that problem, the contrastivist anti-skeptical strategy founders. Without a workable criterion for distinguishing relevant and irrelevant alternatives, there are no grounds available to the contrastivist for denying the relevance of the possibility that I am a BIV.

In sum, the argument above has the following structure. Schaffer claims that one may know that it is a goldfinch rather than a raven, while not knowing that it is not a canary. I argued that this is implausible. Either one can rule out the possibility that it is a canary (or any other bird, and so knows it is a goldfinch, rather than *any other bird*), or one cannot. In either case, the possibility that it is a canary is relevant, though the inquiry-guiding question fails to generate this alternative. Sub-conclusion (1): The contrastivist account of relevant alternatives is false (incomplete). Sub-conclusion (2): With no *other* account of relevant alternatives, the contrastivist cannot distinguish seemingly relevant alternatives (that I was born without hands; that it is a canary) from seemingly irrelevant ones (that I am a BIV; that it is a clever fake goldfinch). Conclusion: Insofar as the BIV possibility may be deemed relevant (given (2)) to whether I know that I have hands, and I cannot rule out that possibility, the threat of widespread skepticism even about our ability to know ordinary truths is raised anew.

Again, we need a better account of relevant alternatives. The modalism I favor delivers this by entailing that the relevant alternatives are those that would be the case in the nearest worlds where the proposition in question is false. I know that I have hands because, in all the nearby worlds where I do not have hands, I do not believe that I do, and, in addition, the cognitive processes through which I form such beliefs produce mostly true beliefs throughout nearby worlds. BIV worlds are distant and so irrelevant. But the question of whether I am not a BIV makes relevant the counter-possibility, though distant, that I am a BIV. In the closest worlds where I am a BIV, I still think I am not. I have no cognitive mechanism for distinguishing BIV from non-BIV worlds. The *safety* theorist may complain that it is unmotivated to appeal to such distant worlds in assessing whether I know I am not a BIV, but here we are comparing theories that feature the ability to distinguish as essential to knowledge, and so we may safely relegate safety to the sidelines.

As we have seen, much of the contrastivist model is designed to capture the ways in which inquiry progresses, where the concern is that, without keeping track of the question that guides the inquiry, there would be no way to gauge progress. This much is surely correct. What I have been arguing is that making progress, for example ruling out "raven" by noting that the bird is yellow, is not itself sufficient for the truth of the claim that it is a goldfinch rather than a raven, precisely because one does not know, by hypotheses, that it is not a canary. Progress is one thing, knowledge another.[33]

I readily admit that this discussion is all too brief, and thus perhaps does injustice to contrastivism. But this is a study of the prospects for modalized epistemology. I have included this digression not with a view toward *refuting* contrastivism, but to acknowledge that it is a formidable alternative theory based on the intuition that knowledge involves a distinguishing capacity, and to offer some of my own reasons for not pursuing it.

F. Conclusion

Though this chapter is ostensibly (and truly) concerned primarily with safety, perhaps the main result I sought is that satisfaction of local reliabilist conditions is by itself insufficient for knowledge. This is a point I made in the introduction, and now I hope its grounds are somewhat clearer.

But we have also seen from Chapter 2 that, while appeal to the idea of a reliable belief-forming process does illuminate the field of knowledge, it does not go far enough. First, on illumination. Modal principles do not say much about how knowledge is achieved. Though process reliabilism is pitched at a fairly high level of generality, it at least locates knowledge where we would pretheoretically expect to find it – in the cognitive processes that produce our beliefs. It should not startle us to find that theories of knowledge that make no mention of our general belief-forming processes are deficient. Second, because process reliabilism is also, taken by itself, an insufficient explication of the requisite belief-truth link constitutive of knowledge, in particular because it does not eliminate veritic luck, we have seen that a correct theory requires that specific beliefs are not merely true, but must be true at least in nearby worlds, too. The remaining issue was which local modal principle best accords with our intuitions about knowledge and eliminates veritic luck.

I have argued that once we fully grasp the need for and nature of process reliabilism, we ought to accept sensitivity, or at least sensitivity rather than safety, as a further condition on knowledge. Specifically, I have argued that safety and process reliabilism are uncomfortable bedfellows. Genuine reliability requires responsiveness to the world, whereas safety is content in some cases to get on without it. Sensitivity, on the other hand, complements process reliability beautifully, since it also demands responsiveness (in the form of discrimination)[34] – if *p* were false, S would not believe that it is true. This speaks strongly in its favor.

But what of the attractive features of safety mentioned in §A? One of them was its superior anti-skeptical results; not only is it compatible with knowledge of everyday truths, it makes space for knowledge of the denials of skeptical hypotheses. I trust that this is no longer compelling. If you think it is, then you have already, for whatever reason, rejected my arguments, and there is little remaining for me to say. The second attractive feature of safety is that it (allegedly) upholds the intuitively persuasive closure principle. But of course, that principle is only persuasive if you still think that we can know that skeptical hypotheses are false, or if you have given up on the possibility of knowledge altogether.

Suppose instead that you are inclined to accept the sensitivity-plus-process-reliability theory as outlined so far. Presumably you have some remaining questions, and they probably sound like these. (1) What are the details of the theory? In the next chapter, I will fill in as many as I deem fit for purposes of having in hand a viable theory, or at least theoretical framework. I will also test it against some of the salient problems already raised, plus a couple of new ones. (2) Even if closure is false, how are we to understand that? Well, its falsity *is* counterintuitive. To claim otherwise is to engage in wishful pretense. However, contextualists have argued for some clear criteria on the adequacy of responses to skepticism, including responses that involve rejection of closure. In the first half of Chapter 6, I argue that reliabilism can satisfy at least those criteria that are unquestionably nonnegotiable. My suggestion will be that, if the theory can explain the deep intuitive pull of skeptical arguments while also adhering to our best understanding of knowledge, then its denial of closure will not appear to be a mere ad hoc maneuver to outflank the skeptic. (3) Why are you averse to agent reliabilism and virtue theories that achieve the same results as modalized process reliabilism? I am not sure that I am, depending on what one means by "agent reliabilism." In the next chapter, however, I offer some considerations against thinking that beliefs based on intellectual virtues (or, for that matter, proper functioning in Plantinga's sense) are the sole candidates for knowledge. In Chapter 6, I take up a challenge from some virtue theorists, namely, that reliabilists cannot account for the added value of knowledge over mere true belief, since for the reliabilist attaining true belief is the only epistemic good. My first inclination is to say that it is at least not obvious that knowledge is more valuable, but even so I think we can locate a source of value in reliably formed beliefs that is independent of attaining the truth. If my reply answers the challenge, it gives yet another reason for not *requiring* that knowledge derives from cognitive or intellectual virtue.

5 Knowledge: Reliably formed sensitive true belief

With much of the groundwork complete, we can now focus on the modalized theory of knowledge rising from the foundation. There will be few surprises in this chapter, as I have already introduced, in the introduction, the view that knowledge is sensitive true belief formed by a reliable process, and since then worked toward confirming that position in Chapters 2–4. The tasks here, then, are to set out the theory, fill in the details that are warranted, distinguish the theory from some of its well-known cousins, and defend the theory in light of previously raised problems and a couple new of ones. Without (still) further ado …

A. The theory

S knows that *p* if and only if:

(1) *p* is true.
(2) S believes that *p*[1].
(3) S's belief that *p* is formed by a belief-forming process or method_w that produces a high ratio of true beliefs in the actual world and throughout close possible worlds (reliability condition).
(4) If *p* were false, S would not believe that *p* via the method_n S actually uses in forming the belief that *p*[2] (sensitivity condition).

I use subscripts to distinguish senses of method in (3) and (4), where "method_w" in (3) refers to the more widely individuated methods discussed by Goldman, which he distinguishes from processes. Methods include "various sorts of algorithms, heuristics, or learnable methodologies," whereas processes are part of our unlearned, typically innate cognitive endowment.[3] "Method_n" in (4) refers to the narrowly individuated methods introduced by Nozick in order to solve, among others, the grandmother case. I shall have much more to say in clarifying the narrow methods versus wide methods and processes distinction below, especially in my solution to the generality problem (§B.4). In the meantime, because the kind of reliabilism that I favor is traditionally called "process reliabilism," I will sometimes just use the word "process" when discussing methods that are widely individuated because talk of methods may create unwanted confusion.

1. Comments on the reliability condition (3)

We saw in Chapter 4 that there are thorny questions concerning the ratio of possible worlds in which one's belief must be true to satisfy safety and thus to constitute knowledge on that view. Need it be true in many, most, nearly all, or all close possible worlds? (This is not a question about the distance of the possible worlds through which one's belief must be true; rather, assuming that issue to be settled, it is a question of how many – the ratio or percentage – of those nearby worlds in which one's belief must be true.) The reliability condition does not raise that problem. The concept of reliability applies to all the outputs of the process, and because it does not require that all of its outputs are true, we should say that reliability requires that the process produce mostly true beliefs in *all* close possible worlds, worlds where any given fact or facts may be different than in the actual world, but where there are no *systematic* differences, for instance in the laws of nature or the appearances of objects generally, that could convert actual-and-nearby-world truth-conduciveness to potential large-scale counterfactual proneness to error.[4]

Reliability entails truth-conduciveness throughout all nearby worlds. We have also been operating under the assumption that reliability requires truth-conduciveness *only* in nearby worlds, and that idea needs at least a brief defense. The following example illustrates that our pretheoretic notion of reliability entails truth-conduciveness throughout nearby worlds but not in distant worlds. I have recently begun playing, or more accurately learning to play, the guitar, and I am getting better at recognizing notes and chords. (I have never had any sort of music education or training, so let's pretend for a minute that I'm pretty good at this.) Do I know a G7 chord when I hear one? (CHORDS) At the end of this chapter, I will argue that processes and methods_w should be individuated very narrowly, but not so narrowly as to include *specific* belief contents (which *will* be included in the individuation of methods_n). For now, let us say that my process is *basing beliefs about tones on listening attentively to the sound from the amplifier.* I know that the chord I hear is a G7 only if I am a reliable listener: generally, only if most of my beliefs formed from that process are true, more specifically, only if I can distinguish G7 from other chords. I need not be infallible, but if I frequently misidentify chords, and in particular if there is some chord that I am prone to mishear and subsequently believe is a G7, then my process is unreliable, and it is plausible to say that I do not know a G7 when I hear one, even in the case where my belief is true.

Now we want to explore the relation between reliability and possible worlds. We saw from the brain lesion and benevolent demon cases in Chapter 2 (§C) that a process that produces true beliefs only in the actual world is not sufficiently reliable to yield knowledge. Suppose that I practice guitar only on Sundays from noon to two p.m., and that my beliefs about chords and notes, based on the process above, are almost always true, hence that in the actual world my belief-forming process is truth-conducive. Similar to the brain lesion and benevolent demon cases, my true beliefs about chords based on careful listening to the sound from the amplifier do not constitute knowledge if they are true *only* on Sundays between noon and two p.m. – if at other times they would be false.

Actually, this last claim needs a little refinement. Suppose that Sundays between noon and two are the only times I take medication for migraines, and thus are the only times I am able to concentrate enough to hear music well. Given the way I will propose to individuate processes toward the end of the chapter – very narrowly but not by reference to specific belief contents (which would be narrower still) – this would constitute a distinct process, for instance, *basing beliefs about tones on listening carefully to the sound from the amplifier when on my medication*, and it *would* be reliable. But here I am trying to illustrate the way in which a process that yields beliefs that are accidentally true does not produce knowledge. Instead of my true beliefs being reliably formed as in the case where I take medication, suppose that I have a demon helper on early Sunday afternoons. Does the existence of the demon helper also constitute, as taking medication does, (part of) a distinct process? If so, why say that the process is unreliable? Odd as it may seem, getting clear on the answers to these questions matters quite a bit. Let me explain.

Suppose that the demon works on my cognitive processes, rather like the medication. That is, the demon affects the way I hear notes and chords, helping me to distinguish them more clearly. Described in this way, and assuming that processes are narrowly but content-neutrally individuated, it is not implausible to say that this constitutes a distinct process, say, *basing beliefs about tones on listening attentively to the sound from the amplifier with the help of the demon.*[5] (The demon is causally operative on my belief-forming processes. I do not have to know that this is the case – to understand how my beliefs are formed – for this to be relevant to a correct description of the process at work, just as I do not have to know that I take migraine medication. Perhaps a friend slips it into my lunch every Sunday right before I practice guitar. Note that this demon case is different than Greco's benevolent demon, where the process used is reasoning according to the gambler's fallacy, and the demon arranges things so that the "predicted" roulette number subsequently comes up. The benevolent demon has no impact on the gambler's cognitive processes.) Individuating the process as above, it does not make sense to locate any presumed unreliability of this process in the nonexistence of the demon in many nearby worlds, for that would be like saying that the medication is unreliable because there are nearby worlds where I fail to take it. (This is to confuse the "unreliability" of the existence of the process with the unreliability of the process itself.

Only the latter is at issue. Pritchard discusses the significance of this.[6]) Rather, if the demon process is unreliable, it is because there are nearby worlds where the demon is as likely to affect my cognitive processes negatively – to make me hear worse – as he is to affect them positively. (Contrast this with my migraine medication, which other things being equal allows me to distinguish chords not only in the actual world but throughout nearby worlds.)

This is the sort of thing I had in mind in saying that my beliefs about chords, true only on early Sunday afternoons, are unreliably formed. Suppose that at other times the demon would hinder my hearing. The demon process is unreliable because it is not truth-conducive throughout nearby worlds (worlds where I practice at other times, too, and the demon confounds my hearing). These are results that we can accept. It follows, on this view, that a demon helper who aids but never impedes my hearing can be part of a source of knowledge, even if he exists only in the actual world and not in many nearby worlds, or exists but does not always intervene – so long as *when* he intervenes I form true beliefs. But let us not be confused about the seemingly embarrassing consequences. Nobody thinks there are demon helpers, so the suggestion that if they existed, they could be sources of knowledge sounds outrageous. But if they did exist, and they did help us, for example, see and hear better, and they did not hinder us, then why deny that they could aid us in knowing?

In any case, for present purposes, you do not have to accept the idea that a demon helper can be part of a reliable process that produces knowledge. The issue is the relation between reliability and possible worlds, and the example shows at least that, if a process, when it is used, does not produce mostly true beliefs in nearby worlds, it is not reliable. Surely the world where I practice guitar at times other than noon to two on Sundays is close to the actual world in which I do not. We know this independently of any discussion of knowledge and reliability. We also think that my beliefs about notes and chords are not reliably formed if the process by which I come to have them – whatever it is – is the process used at many times other than early Sunday afternoons but produces true beliefs only then. We know this independently of any discussion about nearby and distant possible worlds.

Imagine, on the other hand, a possible world where sound waves randomly change in frequency. This world is systematically different, and thus intuitively far more distant, from the actual world than a world where only my practice times are different. In this world, my actual world belief-forming process, whether *basing beliefs about tones on listening attentively to the sound from the amplifier* or *basing beliefs about tones on listening attentively to the sound from the amplifier when on my medication*, does not produce true beliefs about chords and notes. Notice, though, that this sort of world is irrelevant to whether my actual world beliefs are formed by a reliable process, or at least I am wagering that most of us would think so. The upshot is that we have intuitions about reliability that are secure (even if not precise) and can be understood independently of, even though they naturally go together with, our intuitions about which possible worlds are more distant than others. We think of reliability as stability of truth-conducive belief-forming processes, whereas we think of distance as a function of difference from the actual world, where systematic differences constitute Greater distances. It turns out that our beliefs about reliability coincide with our intuitions about the distance of possible worlds in the following way: Greater stability implies truth-conduciveness and responsiveness throughout more worlds, which means greater reliability in the belief-forming process.

I have mentioned several times already that I favor a general reliability condition for knowledge over some of its close relatives, including Plantinga's proper function account and the extant versions of agent reliabilism and virtue theory. Plantinga responds to his brain lesion case by adding a proper function condition on knowledge.[7] Briefly, a cognitive process or faculty yields knowledge only if it is functioning properly, that is, as it was designed, whether, for instance, by its creator or by evolution. But as Greco rightly points out, this is too restrictive. He cites cases (from OLIVER SACKS) where "abnormality and dysfunction increase the victim's cognitive abilities along certain dimensions and, as a result, increase the person's capacity for ... knowledge."[8] One is the case of autistic twins with extraordinary mathematical abilities, another of "a man whose illness resulted in an increase in detail and vividness concerning childhood memory" (Greco 2003a, "Virtue and luck," 357). Surely knowledge is achievable through abnormal or improperly functioning but nonetheless reliable processes, ones that were not in the design plan, no matter how one thinks of design.

Accepting this point requires straightening out one of the concerns regarding process luck. Recall my explication of process luck (from page 15): A belief that ϕ is formed by a lucky process if, and only if ... the process by which it is formed produces a significant ratio of false beliefs in many nearby possible worlds. It may be, in one sense, merely accidental (dare I say "lucky") that the man in the Oliver Sacks case came to be ill in such a way as to produce such vivid memories, but, given that this belief-forming process is stable in the relevant respects, the true beliefs issuing from it are not process lucky, since the process produces mostly true beliefs throughout nearby worlds. It is clear that there will always be *some* element of luck in our achieving knowledge, if only because the belief-forming processes we have are not entirely up to us – we do not choose, for example, our natural birthright – and because it is, in some sense, lucky (if true) that we do not live in the evil demon's world, but for all that we do not think that luck of these sorts precludes knowledge.

I take this last insight also to weaken, at least somewhat, the rationale for versions of agent reliabilism and virtue theories motivated by the desire to capture the element of credit due to an agent when the agent knows, as opposed to lack of credit when her true beliefs are merely lucky. The general idea is that knowledge is incompatible with epistemic luck, and lack of credit due to an agent indicates the presence of epistemic luck, and so knowledge requires credit. I have just argued that some kind of luck – in the form either of the world cooperating with our belief-forming processes or of our having unique abilities, such as the autistic twins', neither of which constitute process or veritic luck, as those notions are defined – is compatible with and perhaps even necessary for knowledge. The question is: Does this kind of luck diminish credit due to an agent in cases of knowledge? If the answer is "yes," then it would seem that knowledge may not entail credit after all.

The clearest notion of credit due to an agent applies when we deem the agent responsible for her holding the beliefs she does.[9] We have long ago set aside these strongest, responsibilist theories of credit cum justification, for they tend toward internalism, which, I have claimed, may be a crucial feature of the traditional notion of justification, but is not necessary for knowledge.[10]

One of the best expositions of credit due to an agent that does not entail responsibilism and is firmly in the externalist tradition is Greco's agent reliabilism. Greco offers analogies from the world of sports to show that we sometimes give credit to an athlete while withholding it from another athlete who performs the same feat, because the former demonstrates a virtue that, by definition, is stable, that is, reliable. We can imagine that such virtues are not the result of constant practice and the pursuit of perfection, but are innate. Even so, Greco claims, we give credit to an athlete for a good performance when it is "in character," but not when it is merely lucky.[11] Greco wants to include both innate and cultivated belief-forming mechanisms as aspects of cognitive character:

> Accordingly, innate vision gives rise to knowledge if it is reliably accurate. But so can acquired skills of perception and acquired methods of inquiry, including those involving highly specialized training or even advanced technology. So long as such habits are both stable and successful, they make up the kind of character that gives rise to knowledge.
> (Greco 2000, 177)

Credit (and, presumably, blame) is attributable to an agent based on her character; some features of character are innate; so credit accrues to an agent whose reliably formed beliefs are based on innate processes.

Or maybe not. I do not want to claim that Greco is wrong, but I do think this is an open question. The fact that we admire an athlete with native skills does not mean that we give her extra credit. Sometimes we do just the opposite – we complain that she was just lucky to have that native endowment. "*I* could dunk a basketball if I were six-foot-six!" Again, the kind of luck due to native endowment is compatible with knowledge, in that beliefs formed from reliable innate processes are themselves not accidentally or merely luckily true, though one may be lucky to be so blessed by nature.

(Compare John Rawls on a similar issue. His description of the original position, particularly the veil of ignorance, from which rational parties hypothetically choose principles of justice, is meant to achieve an outcome that "is not conditioned by arbitrary contingencies."[12] Thus, from the original position, claims Rawls, one does not "know his fortune in the distribution of natural assets and abilities, his intelligence and strength, and the like" (Rawls 1971, 137). The implication is that these are not morally relevant. Rawls has been criticized for this, but his view at least casts doubt on the notion that agents deserve credit for their innate abilities.) When someone – an athlete or inquirer – does something to improve her athletic or belief-forming capabilities, we do indeed give credit for achievements based on those abilities. But the autistic twins and illness-induced enhanced memory cases are different, and maybe we do not give those agents credit. We may still think highly of them, and even wish we had such capacities for knowledge, but whether we give them credit is unclear. I worry that we are led down the path of responsibilist, hence internalist, justification[13] if our aim is to account for credit due to an agent when she knows, if only because "credit" is a vague term, or, as I shall suggest in the next chapter, it is used ambiguously. Our process reliabilist theory can do everything Greco's agent reliabilism can do, and even more, once we include the sensitivity requirement, with one possible exception. Eschewing the demand to include credit due to an agent, we are faced with the value problem for reliabilism – the worry that it will be impossible within our chosen theoretical framework to explain the added value of knowledge over mere true belief. I will take this up in the final chapter.

In the meantime, I do not want to leave you with the impression that the foregoing comments sufficiently clarify the reliability condition. There remains the looming generality problem for reliabilism. Solving that problem will require very carefully distinguishing processes and methods_w from methods_n – the latter being included in the sensitivity condition. I first want to clarify sensitivity as much as possible, and so I will postpone treatment of generality and other remaining problems until after the sensitivity discussion.

2. *Comments on the sensitivity condition (4)*

The main question here is one already discussed in Chapter 2, namely, which possible worlds are relevant to the truth conditions of the sensitivity conditional? A version of the concern lurking implicitly in this question came up in the introduction – that God only knows which possible worlds are relevant, and so the very idea of appealing to possible worlds to clarify what would or might be the case is like letting the blind lead the blind.

Before settling the main question, let us note again that, though we may not always have a precise ordering of possible worlds, we have a pretty clear understanding of distance. Recall the example from the introduction. I know that in the closest possible worlds where I am a millionaire: (1) I have more money than I do now; (2) some people ask me for money occasionally; (3) I invest at least some of it; (4) I give at least some to charity; (5) I eat well; and (6) I travel more. That is, I know some things that would be the case if one of those worlds were actual. Intuitively, then, I know that those worlds are closer, less different, than worlds where I am a millionaire and I wallpaper my office with ten-dollar bills.

So which close possible worlds are relevant to sensitivity? Maybe God only knows, but, I want to say, that's okay. In Chapter 2 I offered two possible interpretations of sensitivity in response to Goldman's thermometer counterexample, and I claimed that, perhaps precisely because God only knows, we should be willing to accept either of them. The first was what Goldman (1986, 45) calls the "pure subjunctive" view, where only what is the case in the closest not-*p* world is relevant to assessing sensitivity. On that view, it is true, in the thermometer example, that if the child's temperature were not 98.6 degrees, the parent would not believe that it is. The closest world where it is false that the child's temperature is 98.6 degrees is still one where the parent uses the one accurate thermometer, rather than one of the inaccurate ones. Is this the right construal of the subjunctive conditional? Frankly, I'm not sure. But, to the very extent of my uncertainty, I am also not confident in claiming that the parent does not know the temperature. One way to account for the indecision is to say that the parent knows the child's temperature, but just barely. She satisfies the minimum requirement. Her epistemic position is just strong enough, just sensitive enough,[14] to cross the lower threshold necessary for knowledge, but it could be stronger.

On another interpretation of the subjunctives, one that I claimed Nozick seems to have had in mind, one's belief that *p* satisfies sensitivity only if it tracks the truth or falsity of *p* not just out to the nearest not-*p* world, but at least some distance further. On this reading, the parent's belief violates sensitivity, because there is a relevant not-*p* world where she nonetheless believes that *p*, that the child's temperature is 98.6 degrees. This is a world where not only is the child's temperature different than in the actual world, but the parent also unwittingly selects a broken thermometer (which happens to be stuck on 98.6 degrees). This kind of example shows that Nozick is wrong to think that his subjunctive account can clarify, better than relevant alternatives accounts themselves, which alternatives, that is, possible worlds, are relevant, because we do not understand the truth conditions of sensitivity without a prior understanding of what is relevant, hence the "God only knows" objection.[15] I think that Nozick himself must have been attuned to this problem when, in the crucial footnote, he says, "many *p* worlds that interest us may have their distances from [the actual world] matched by not-*p* worlds" (Nozick 1981, 681; note 8). He is explicitly exploiting our sense of relevance, of the "worlds that interest us," in formulating his interpretation of subjunctives. But I do not see the lack of a perfectly precise rendering of the possible worlds relevant to assessment of subjunctive conditionals as a problem, at least not for epistemology. Again, the reason is that our intuitions regarding the parent's knowledge status do not provide clear-cut answers, and so we should have different possible interpretations of the conditions requisite for knowledge ready to hand, depending on which way we are leaning. Aristotle's admonition not to seek precision where none is to be found is as relevant to epistemology as it was, for him, to ethics: "Our discussion will be adequate if it has as much clearness as the subject-matter admits of; for precision is not to be sought for alike in all discussions, any more than in all the products of the crafts."[16]

I opened this book with the claim that one of the recent developments making possible a true theory of knowledge is modalism, which I glossed as free use of possible worlds talk in explicating notions of necessity, possibility, and what might or would be the case. I did not claim that we should seek exactness here, and that is one reason why modalism is not only *not* ill suited to our pursuits, but is rather entirely appropriate. Knowledge itself is an inexact notion. We ought to aspire to giving clearly articulated necessary and sufficient conditions for knowledge, but it was never in the bargain that those conditions, taken individually, would or should not admit of vagueness. So what's the use of modalism? Though we may not have a clear-cut understanding of which possible worlds are relevant to the assessment of subjunctive and counterfactual conditionals, we do have a pretty good, shared understanding of the space of possible worlds, and it is the latter that carries the burden. Throughout this book we have been discussing examples involving possible worlds talk, and I daresay that we could not have gotten this far without it, imprecise though it may be.

My final word on this issue, then, is that we neither need nor should seek a definitive interpretation of the possible worlds relevant to the sensitivity conditional. This is consistent with claiming that there is something deeply right about the idea that a true belief does not constitute knowledge if, were the proposition believed false, the agent would believe it anyway. Indeed, I argued in Chapter 4, against safety, that this fundamental point dovetails nicely with a reliabilist outlook, hence that conditions (3) and (4) are complementary. But if you absolutely insist on an exact criterion for sensitivity, I offer the pure subjunctive view, according to which minimal knowledge requires that, in the nearest possible world where one's belief is false, one does not believe it.

B. Problems and clarifications

Rather than filling in the remaining details directly, I shall further explicate the theory endorsed above by considering its responsiveness to a series of problems. I first briefly revisit some of the issues that have already been resolved. In §2, I introduce a seemingly odd consequence of the theory that is closely related to its incompatibility with the closure principle. We then turn to two outstanding issues raised back in Chapter 2, the serendipitous reliable process problem (§3) and the generality problem (§4).

1. Some old problems revisited

If it is not already apparent, there are two giants upon whose shoulders I stand in setting out the theory of knowledge as reliably formed sensitive true belief – Goldman and Nozick. But having rejected Nozick's fourth condition – if *p* were true, S would believe it – there is nothing left in his theory to account for knowledge of necessary truths. As we have seen, the antecedent of the sensitivity conditional is never satisfied for such truths, making it difficult at best to understand how S's belief in a necessarily true proposition might be truly sensitive. (In addition, we saw too from the Fermat's Last Theorem problem that satisfying Nozick's fourth condition is insufficient for knowledge of necessary truths anyway.)

Well, that's the nice thing about having *two* giants to support me. We can say that, for beliefs in necessary truths, the sensitivity condition is always trivially satisfied, since there is no possible world that would make the antecedent true, in which case the conditional is itself true. (This is Lewis's view, which he admits is controversial.[17] Perhaps subjunctive conditionals do not behave anything like the material conditional, which is true when its antecedent is false. Of course, I am not saying precisely *that* about the subjunctive; I am suggesting that it may be trivially true if the antecedent is *necessarily* false. If that seems wrong, we can fall back on Nozick's claim that "condition 3 does not come into play for necessary truths" (Nozick 1981, 186) and seek some other necessary condition for knowledge of necessary truths, which is forced upon us in any case.) And then, in accordance with our theory, we will say that knowledge requires not just that one's true belief is sensitive, but also that it is reliably formed. In the Fermat's Last Theorem problem, we imagined that a mathematician believes the theorem to be true because no counterexample has been discovered (and we supposed also that it is in fact true). His belief satisfies Nozick's fourth condition, for (we can stipulate that) he believes it in all nearby worlds, but his belief-forming method_{w} is unreliable (just as Garrett says).[18] While the mathematician's belief is true in the actual world, the method would produce false beliefs in many nearby worlds.[19]

The two other categories of problems attaching to the sensitivity-based account of knowledge are that it is too permissive and that it is too restrictive (though the problem about knowledge of necessities could be filed under the former). These were treated in Chapter 3, so in the interest of avoiding excessive repetition, I shall be brief.

Let's look at a case representing the first category: Luper-Foy's case involving Flip.[20] Recall that Flip bases his beliefs about whether the light is on in the next room on a coin toss. However, Jessica is watching Flip, and when the coin comes up heads, turns the light on, and when tails turns it off. So if it were false that the light is on, Flip would not believe it. But surely Flip does not know that the light is on when it is. In Chapter 3 I said that adding a process reliability requirement can help. The process Flip uses is basing belief on a coin toss, and that general process is utterly unreliable, hence lucky when from it one forms a true belief, which is why Flip does not know. It is unreliable because, though it consistently produces true beliefs in the actual world, there are many nearby worlds where it produces false beliefs, for instance, those where Jessica is not around, or gets bored with her little game, or whatever.

Back in Chapter 3 I argued that a process reliability condition can also produce a nice resolution of the Haveit case.[21] Haveit wants me to believe truly that someone in the office owns a Ford, so he produces false evidence that Nogot owns one and withholds evidence that he himself owns a Ford. If it were false that someone in the office owns a Ford, I would not believe it because Haveit would not have given evidence that Nogot owns one, but I do not know despite my sensitive belief.

It is doubtful that the belief-forming process is reliable, however. How precisely to understand the nature of the unreliability is a question we cannot fully answer until we resolve the generality problem, below. Relying on testimony in general might very well be deemed reliable, being truth-conducive in the actual world and in nearby worlds, in which case the result would be that my belief is both sensitive and reliably formed, but not knowledge. What we need is access to a nonarbitrary criterion of process individuation that allows for narrower process types. One important constraint on more finely individuating process types is that we do not collapse the processes and methods$_w$ versus methods$_n$ distinction, which would threaten to undermine the distinction between the reliability and sensitivity conditions. The solution I will offer to the generality problem satisfies this constraint.

With respect to the second category of complaint with the sensitivity-based account of knowledge – that it is too restrictive – the main problems are induction and higher-level knowledge. I offered an extended treatment of the latter problem in the sensitivity chapter; thus, for details, I refer the reader to Chapter 3, §C.3. In a nutshell, the problem is that, were my higher-level belief (that my first-order belief is true, or not false) itself false, I would nonetheless think it was true. My reconciliation of sensitivity with knowledge that one's first-order belief that *p* is true (or not false) required that we distinguish two kinds of higher-level propositions: those that do not entail that one actually has the first-order belief (Vogel's); and those that do (the ones I introduced). I argued (a) that the latter are compatible with sensitivity because nearby worlds where they are false are typically ones where one does *not* hold the first-order belief, and thus would not believe that one holds it; (b) that the latter are more natural renderings of the kind of natural language higher-level beliefs that interest us and Vogel, those that involve first-order knowledge; hence that (c) Vogel's argument is not very damaging at all. Vogel's argument that sensitivity undermines higher-level knowledge is driven by higher-level propositions such that, in the nearest possible world where my actual world first-order belief is false, I *still hold* that belief, and thus have a false higher-level belief that my first-order belief is not false. But if my first-order belief is in fact a case of knowledge, then the world where I hold that belief but it is false must be more distant than the worlds relevant to whether my first-order belief is sensitive. That is, for such propositions my higher-level belief must be *more* sensitive than my first-order belief. It is not surprising that this kind of higher-level belief could violate sensitivity and thus not constitute knowledge.

The other problem meant to illustrate the overly restrictive nature of sensitivity is its putative incompatibility with inductive knowledge. In Sosa's trash chute case, he claimed that, if it were false that the bag will soon be in the basement, I would believe it anyway. As with the thermometer case, I offered two distinct solutions, the first of which might be considered pretty radical. It is that the kind of knowledge to which we are entitled in drawing inductive inferences is at best probabilistic: I know that it is highly likely that the bag will soon be in the basement, but not that it will be. For an ordinary chute, presumably the kind Sosa has in mind, though there may be a nearby world where the bag does not fall to the basement, it is still highly likely that it will, given the way the chute is designed. A world where it is false that it is highly likely that the bag will soon be in the basement is one where the chute itself is somehow different than it actually is, for example, defective, and it is at least plausible to say that I would notice this and thus not believe that it is highly likely that the bag will soon drop, and if I did not notice it was defective, then of course I do not know even that it is highly likely that it will drop (because this is false).

I rather like this suggestion, but I see too that many readers will think it extreme. In my original discussion of this case in Chapter 3, I argued that it is possible for an agent's belief to satisfy sensitivity in cases of inductive knowledge, provided that the method$_n$ includes not only the positive instances that ground the inference, but also lack of negative instances and lack of evidence of their likelihood. This is consistent with Nozick's criterion of method$_n$ individuation as "set by the differences the person would notice" (Nozick 1981, 233). If the agent would not have noticed any evidence that, for instance, the bag would not drop, then she does not know that it will, but if she is attuned to any possible such evidence (and there is not any), she does know. These constitute two different possible methods, whereas Nozick rightly claimed that the method$_n$ must be held constant in assessing the sensitivity conditional. In contrast, when we are looking at the case where the agent *would have* noticed evidence that the bag will not drop even though, in actuality, there is no such evidence, we are not considering two distinct methods – one where there is only positive evidence and another where there is also negative evidence. Rather, there is just one method$_n$ that allows for and covers different possible experiences. (This is yet another reason why Nozick ought not to demand that one's actual world experiences themselves need to be held constant to count as the same method applied in other possible worlds. See Chapter 3, note 18.) It is because $methods_n$, as opposed to $methods_w$ and processes, are so narrowly individuated – by what the agent would notice as it bears on the specific belief in question – that we can include the specific possible experiences of an agent in them, and thus make sense of the satisfaction of sensitivity in cases of genuine inductive knowledge.

That concludes the discussion of problems previously raised and, to greater and lesser extents, previously treated. Let us turn to a new problem, and to two old problems left unresolved.

2. A seemingly odd consequence

As promised more than once already, I will treat the failure of closure entailed by the theory advanced here in Chapter 6. Meantime, there are cases of closure failure that appear to follow from the reliably formed sensitive true belief theory that seem especially counterintuitive. Fumerton describes one (CONJUNCTION):

> For similar reasons, it also follows on Nozick's account (and this sounds really strange) that I can know the conjunction, that there is a table before me and that I am not being deceived with respect to its existence by an evil demon, even though I cannot know the conjunct, that I'm not being deceived by an evil demon. Intuitively, the close possible worlds in which the conjunction is false are the worlds in which it is false that the table is there, and these, by assumption, are worlds in which I would not believe that the conjunction holds. The close possible worlds in which the demon hypothesis is true are very far removed from the actual world, and in these worlds we would continue to believe that the demon hypothesis is false.[22]

Strange indeed. This is an instance of closure failure because the conjunction, which seems to satisfy sensitivity and is therefore known, clearly entails each of its conjuncts, one of which is, on Nozick's theory, unknowable. Fumerton is correct to point this out, and I have no objection to his description of the relevant possible worlds here. However, I do not think this is a very troublesome example because, unlike other cases of closure failure, neither Nozick nor I need to accept this one. There's a quick fix.

The right thing to say here is that the theory proffered above is a theory of knowledge for non-conjunctive propositions. If S knows that there is a table in front of her but does not know that she is not being deceived by an evil demon, then S does not know *both* that there is a table in front of her *and* that she is not being so deceived. Already this constitutes a case of closure failure (when properly laid out), but not of failure of knowledge for conjunction simplification. We can then add, what is plausible anyway, that a necessary condition for knowledge that a conjunctive truth-functional proposition is true is that one's knowledge "accords with" the truth table defining the truth conditions for conjunction. Thus S knows that *p* and *q* only if both S knows that *p* and S knows that *q*. (Things get messy, and implausible, pretty fast if we try to generalize the strategy to other kinds of truth-functional propositions, so I won't do that. Surely one can know that either *p* or *q* without knowing that *p* or knowing that *q*.)

Fumerton is right that in the nearest world where the conjunction is false, I do not believe it, but wrong to think that Nozick and his followers are committed to saying that one can therefore know the conjunction while not knowing one of its conjuncts. Rather, what we can and should say is that the theory is designed to apply to non-conjunctive propositions. One might complain that this move is ad hoc, since it does not generalize to the other truth-functional operators. But clearly conjunction is a special case. *S knows that p and q* says that S knows that two distinct propositions are true, unlike *S knows that if p then q* or *S knows that either p or q*. All I am doing is squaring this simple fact with the sensitivity-based account.[23]

3. Serendipitous reliable process

In Chapter 2 (§C) I introduced the serendipitous reliable process problem. Suppose a true believer forms her belief by a method_w that is in fact reliable – truth-conducive in the actual world and throughout nearby worlds – but uses the method_w only haphazardly. Greco offers the case of the CARELESS MATH STUDENT:

> Suppose that S is taking a math test and adopts a correct algorithm for solving a problem. But suppose that S has no understanding that the algorithm is the correct one to use for this problem. Rather, S chooses it on a whim and could just as well have chosen one that is incorrect. By hypothesis, the algorithm is the right one, and so using it to solve the problem constitutes a reliable process. It seems wrong, however, to say that S thereby knows the answer to the problem.
>
> (Greco 2003a, "Virtue and luck," 354)

Greco calls this a fleeting process and uses the example to motivate his agent reliabilist theory:

> The problem is not that the algorithm is unreliable – it is perfectly reliable. The problem, rather, is that the *student* is not reliable – his use of the algorithm is not grounded in a cognitive ability, and so there are close possible worlds where he chooses the wrong algorithm instead of the right one.
>
> (Greco 2003a, "Virtue and luck," 356; emphasis in original to highlight the *agent's* defect.)

This is a nice example. It allows me to say again, but in a different way, why I am not inclined toward agent reliabilism or virtue theory insofar as the rationale for those views is to account for credit due to an agent, and it provides an occasion to clarify the nature of the reliabilism that I favor. To both of those ends, let me contrast the careless math student with the NATURALLY GIFTED MATH STUDENT. The naturally gifted math student, an example of whom may be one of the autistic twins discussed above (§A.1), is like the careless math student in that she "has no understanding that the algorithm is the correct one to use for this problem" (Greco 2003a, "Virtue and luck," 354). But the naturally gifted math student's process is not fleeting, at least not in the sense that she might have selected the wrong algorithm and from it she might easily have formed false beliefs. Instead, she always gets the right answer, because she always uses correct procedures and algorithms, though she has no idea how or why she chooses the algorithms she does. The process (or method$_w$) used by the careless math student seems to be fleeting, whereas those used by the naturally gifted math student are not.

To explain this, Greco describes the situation differently than I would. He claims that the process is reliable, but that the agent is not. But surely there is an alternative description that is at least as compelling, regardless of one's theoretical commitments: The careless math student and the naturally gifted math student use the same algorithm for the problem at issue, but *not* the same process (or method$_w$). The former chooses the algorithm on a whim: "his use of the algorithm is not grounded in a cognitive ability" (Greco 2003a, 356). Surely the difference between choosing an algorithm on a whim and choosing it from a cognitive ability, even one that is not at all understood by the person who exercises it, constitutes a difference in belief-forming process. Recognizing this suffices to solve the serendipitous reliable process problem; that is, we need only recognize that the careless math student's belief-forming process is not reliable.

Consider an analogy. Suppose that Farmer Smith and Farmer Jones are both looking to hire a chicken sexer (CHICKEN SEXER CHOOSING). Smith picks a person at random who happens to be a chicken sexer, and he forms true beliefs about the sex of chicks based on the reports from his employee. Jones, you might have guessed, is a chicken sexer detector. He forms (modally reliable) true beliefs about who can and who cannot determine the sex of chicks, though he has no clue how he does it. He too forms true beliefs about the sex of chicks based on reports from his chicken sexer. Both Smith and Jones utilize reliable chicken sexers, but it does not follow that both use reliable belief-forming processes to determine the sex of chickens. We could at this point distinguish first-order and second-order belief-forming processes to describe this situation and that of the math students, but things are simpler if we think of higher-order processes as implicated in the correct descriptions of the entire processes an agent uses.

What, then, does constitute the difference between the process used by the careless math student (or Farmer Smith) and that used by the naturally gifted math student (or Farmer Jones)? I introduced the naturally gifted math student by saying she was relevantly *similar* to the careless student. Neither has any understanding of why the algorithm is chosen. But, by hypothesis, the careless student chooses it merely on a whim, which is akin to forming beliefs based on a coin toss (like Flip) or basing beliefs on wishful thinking. In contrast, the naturally gifted student does not choose the algorithm on a whim. (I suppose it may, in some sense, *seem* like a whim to her, though I take it the experience of choosing is much more like a tug.) Like the chicken sexer (or the chicken sexer detector!), she has an ability, something that is not fleeting and not merely whimsical.

In sum, then, I want to say that Greco is wrong to call the careless math student's belief-forming process reliable. That claim is based on overlooking an important part of the process actually used. And to be entirely explicit, the point is that this explains why the careless math student does not know the answer to the math problem, whereas the naturally gifted math student does.

Is it legitimate to fold the agent's basis for using a method_w or process into the process itself? One might think that these – the process used and one's basis for using it – should be kept separate. The distinction might strike you as a natural one.[24] But is it compulsory? The naturally gifted math student has an ability that the careless student lacks, but what is that ability, other than (part of) some process by which she can do math calculations correctly? I have no further justification for including one's basis for using a method_w in one's process, except for its theoretical utility in getting cases like the serendipitous reliable process right and, more importantly, in solving the generality problem. My solution to the generality problem involves individuating processes in the narrowest, that is, most refined, but still content-neutral way. According to this criterion of process individuation, we can distinguish the processes used by the two math students. Thus part of the justification for folding the basis upon which one uses a method_w into the process used hinges upon the plausibility of my solution to the generality problem, and if it works, then I want to say that the onus is on someone who believes that the basis upon which one chooses a method_w needs to be kept distinct from the process at issue to show that this must be so.[25]

I said also that my introduction of the naturally gifted math student would allow me to explain a bit further why I reject agent reliabilist and virtue-based accounts. It is at least implicit in Greco's description of the careless math student example that the *student* is unreliable because he does not *understand* what he is doing. This may simply be a feature of the chosen example, but it is apt to mislead, for it is then easy to think that cases of knowledge, those where we deem an agent reliable, are precisely those where she is due some credit in virtue of her understanding. Thinking along these lines, we are prone to conclude that lack of knowledge implies that the agent is unreliable, which is to be explained by our unwillingness to give credit to an agent in virtue of her lack of understanding. But we need to see that knowledge need imply neither understanding nor credit. What the naturally gifted math student example shows is that one may very well lack understanding of the processes she uses, that those processes may nonetheless be reliable, hence knowledge producing, and yet whether the agent deserves credit is dubious. After all, she lacks understanding – her having the cognitive abilities that she does may itself be a matter of luck.[26]

This, finally, is why I favor process reliabilism over agent reliabilism. There is, as I see it, precisely one advantage of agent reliabilism, which is its aiming and ability to account for credit, and only that motivates describing the careless math student as an unreliable agent (which is, of course, true) *rather than* as an agent who uses an unreliable process (which is also true). The case of the naturally gifted math student, who knows but does not understand how, and thus whose creditworthiness is doubtful, undermines the primary motivation for agent reliabilism. Fortunately, we can solve the serendipitous reliable process problem by appeal to a more adequate notion of reliable processes.

4. The generality problem

We have one more challenge to overcome, arguably the most difficult for process reliabilism. Here, in a nutshell, is the generality problem for reliabilism: "Without a specification of the relevant [process] type, process reliabilism is radically incomplete."[27] Conee and Feldman think of the generality problem as one that afflicts a reliabilist account of *justification*, but since I have claimed several times that I have little interest here in explicating that notion, it is better to view the problem as applicable to an account of the belief-truth link (my conditions (3) and (4)), whatever we call it, that constitutes knowledge. (You are free to label that link "justification" if you like, but, as I have said in both the introduction and Chapter 2, dangers lurk therein.)

According to Conee and Feldman, the criterion for process types must be principled, for "it is easy to make *ad hoc* case-by-case selections of types that match our intuitions" (Conee and Feldman 2003, 297).[28] Every belief is formed from a process that instantiates countless types. Which one is relevant to assessment of reliability? The criterion I shall offer is that the relevant process type is the narrowest specific-content-neutral one that is causally operative in belief formation.[29] The requirement of specific-content-neutrality is the key to avoiding the charge that the criterion is ad hoc.

It will help our understanding of how this proposal works in solving the generality problem to see how other proposed solutions have failed. Conee and Feldman's paper is very useful for this.[30] They begin with a difficulty for Goldman's (1979, 9–10) suggestion that process types be individuated in a common sense way. First, "there are far too many common sense types to provide a unique identification of the relevant type for each process token … [for example] visual process, perceptual process, tree-identifying process, daytime process, indoor process" (Conee and Feldman 2003, 299) and so forth. Second, "not all beliefs resulting from any one such type are even approximately equally justified" (Conee and Feldman 2003, 299). For example, the belief that there is a tree in the yard based on a brief and hasty scan may be justified, but the belief that there is a sparrow in the tree based on a brief and hasty scan may not be.

Another solution Conee and Feldman assess holds that the relevant process type is "the natural psychological kind corresponding to the function that is actually operative in the formation of belief" (Conee and Feldman 2003, 301).[31] The problem, again, is that every belief token instantiates numerous psychological kinds, hence the relevant process cannot be identified just by appeal to those natural kinds. William Alston further refines his strategy by saying that the relevant processes are "maximally specific" (Alston 1995, 26), which echoes one of the key ideas from Goldman: "the critical type is the *narrowest* type that is *causally operative* in producing the belief token in question" (1986, 50). In reply, Conee and Feldman argue that this has the awkward consequence that "the relevant type is settled by the mere identity of belief" (Conee and Feldman 2003, 302), such that, if the belief is true, the process is perfectly reliable. This is true "on the reasonable assumption that the content of any normally formed belief is causally determined by its antecedent psychological causes" (Conee and Feldman 2003, 302). Feldman calls this "the single case problem."[32] Suppose Smith sees a particular animal and believes truly: That is a dog. The narrowest type that is causally operative might be: If you see something that looks exactly like that in circumstances exactly like this, conclude that it is a dog. (It is important to note that Smith may not apply this process consciously or have any clue that this is what she is doing.) Since Smith's belief is true, the process is perfectly reliable – there are no other appearances and circumstances exactly like these. (More importantly, even if there were such exact circumstances, if we take "that" as a spatiotemporally indexed demonstrative constituting part of the belief content, we get Conee and Feldman's worry that process type "is settled by the mere identity of belief" – the single case problem.) One problem here is that the present proposal for process individuation utterly lacks generality and does not answer to the intuitive idea of a reliable process. Relatedly, though the process turns out to be perfectly reliable (maximally truth-conducive), Smith's process may nonetheless be intuitively unreliable according to what I previously claimed is the fundamental sense of reliability, namely, the capacity to distinguish. Consistently with perfect "reliability" of the narrowly defined process above, Smith might call *everything* else or *nothing* else that she sees a dog!

We generate the same kind of criticism if we try to save process reliabilism by applying this proposed criterion of process types to the Mary problem. Mary's vision is reliable, even in somewhat dim lighting, but she does not know that her husband is in the living room, though it is true that he is and she believes it. The reliabilist could solve this problem, that is, explain why Mary does not know, by implicating the specific content of Mary's belief in the relevant process type, such that her belief-forming process is: *Believe that my husband is now in the room on the basis of a quick look in dim lighting if I see a person whose appearance has such-and-such features.* By hypothesis, Mary's vision is reliable in dim lighting and even when she quickly glances, but perhaps in this one case unreliable when the specific belief content involves her husband on this occasion, since there are nearby worlds, including the actual world, where she mistakes her husband's brother (who sometimes comes around) for her husband in those conditions. Here again, though, appealing to specific belief content in individuating the process type to solve the Mary problem seems utterly ad hoc, not at all general, and thus not in the spirit of *process* reliabilism.

The solution seems, to me anyway, fairly obvious – the processes must be individuated in a specific-content-neutral way. This was Goldman's proposal (1979, 12) before he offered his narrowest, causally operative process type criterion.[33] This completely avoids the ad hoc worry above, and is completely general. Process reliabilists may have turned a blind eye to it precisely because it is impotent to solve cases like the Mary problem, but that is where sensitivity, involving content-specific methods_n, comes in. Mary's process, individuated in a content-neutral way, is reliable: *forming beliefs about which people are in the vicinity based on quick looks in at least dim lighting*. Content-neutrality certainly does not entail absence of any categories, or even particular individuals, in specifying process types. The process used in Mary forming the belief that her husband is in the room may involve the ability to detect that certain features indicate the presence of a human being, or even just her husband – she may utilize a general, reliable, husband-indicating process; the belief that there is a dog in one's sight may be formed by a process that involves general dog feature detection and the ability to distinguish such features; and beliefs about dachshunds may be formed by a process that involves distinguishing certain dog features from others. For instance, Oscar forms the belief that Dach is a dachshund based partly on his ability to distinguish floppy ears from pointed ones – part of a process or method_w. He forms the belief that *that* is a dachshund in particular by his method_n, which says: If what I am looking at now has short legs and floppy ears (and such and so other features), it is a dachshund.

Conee and Feldman's criticism based on lack of generality – the single case problem – is that individuating processes to match our intuitions leads to the potential problem of specifying processes that are applicable only to single beliefs (where, I take it, the worry is about belief tokens). But even if we narrow Mary's process to include reference to her husband – forming beliefs about the presence of my husband based on quick looks in at least dim lighting – the criticism does not apply. The reason is that the process *is* general, repeatable in many possible situations. Methods_n, on the other hand, are individuated by reference to the specific content of a particular belief (believing that that person, here, now, is my husband, based on such and so visual cues), and in the imagined case, the belief based on that

$method_n$ is insensitive. I shall say more below to justify this very narrow but specific-content-neutral individuation of $methods_w$ and processes.

So Mary's narrowly individuated but content-neutral process is reliable, but her belief is insensitive, because if it were false, she would still believe it based on her actual $method_n$: *Believe that my husband is in the room on the basis of a quick look in dim lighting if I see a person whose appearance has such and so features.* (Nobody said that *methods$_n$* have to be specific-content-neutral; indeed, our discussion of them in Chapter 3, including the grandmother, Dach, and Jesse James cases, showed that they could include specific content. Now I am arguing that they should.) We now have the desired result – a criterion for process types that is not ad hoc, solves the generality problem and, when coupled with sensitivity, provides a well-motivated, very powerful theory of knowledge.

An honest question at this point is: Is it legitimate to individuate processes so narrowly? The fact that people have a wide range of different knowledge-producing cognitive abilities suggests that it is. Suppose that Smith has very good eyesight, but only in typical or somewhat dim lighting (say, in a brighter than *very* dimly lit room and between dawn and dusk, inclusive). Smith forms the belief that there is a bird in the tree just before dusk. Her belief is reliably formed from the process of *forming beliefs based on visual appearances of animals with such and so features in typical-to-somewhat-dim lighting*, and if true and sensitive, she knows that there is a bird in the tree. If it were nighttime, when she is prone to making visual mistakes, even if her belief were true, it would not be reliably formed. At night, the narrow description of her process is not *forming beliefs based on visual appearances of animals with such and so features in typical-to-somewhat-dim lighting*, but *forming beliefs based on visual appearances of animals with such and so features in dim lighting*, which is unreliable. This criterion of process individuation accords with our intuitions about the situations in which Smith knows, which is clearly in its favor. This and the following examples also illustrate my claim that process types can and should be individuated in such a way as to account for an agent's particular cognitive abilities. (This reveals how process and agent versions of reliabilism are close relatives. As I said in my discussion of agent reliabilism, the main complaint I have against that theory is that it is motivated by the felt need to account for credit due to a knowing agent, whereas in fact an agent may have cognitive abilities for which, it is arguable, she is not creditworthy.)

Now, suppose that Jones's vision is worse than Smith's – that is, reliable only in a narrow range of circumstances. Surely it goes too far to say that Jones cannot have vision-based knowledge. Jones does not see as well in somewhat dim light, and she is a bit nearsighted. Her vision-based beliefs are truth-conducive only in typical lighting and at close range. Insofar as it makes sense to say that she knows there is a bird in the tree when it is noon and the bird is within, say, 25 feet, it is true that her belief is reliably formed, where the process is something like *forming beliefs based on visual appearances of animals with such and so features at close range in at least typical (but not dim) lighting*.

Brown's vision is worse still. She needs very good lighting, and can make out even middle-sized objects only if they are within a few feet. Does it follow that there is no visual process that is knowledge producing for Brown? Surely not. She knows there is a bird in the tree when she sees one at noon and she is standing right under it. Were she to believe truly that there is a bird in the tree 20 feet away, her belief would only be luckily true because the process is unreliable (*forming beliefs based on visual appearances of animals with such and so features at 20 feet or more*), and she would not know there is a bird in the tree. The more specific process she uses when the bird is much closer – *forming beliefs based on visual appearances of animals with such and so features in better than typical lighting at very close range* – is reliable. The process is more specific, narrower, in that its range of application is more tightly circumscribed. Notice, too, that this set of examples reveals why appeal to generic normal or good conditions is useless, since such conditions must be relativized to the agent in question, where what comprises good conditions for one agent may constitute less than good conditions for another.

For a final illustration of perceptual processes, let's look at a case where a process is maximally narrowly individuated. Suppose White is an oriole savant – she knows nothing about any other kind of animal and indeed cannot tell any other animals or birds apart, but she can tell which creatures are orioles. I want to say that White has a general, reliable process for forming beliefs about the presence of orioles, for example *forming beliefs about orioles based on vision within such and so distance in at least such and so lighting*. "But surely the reference to orioles in the process type is illicit!" Is it? Even on this very narrow individuation, we (1) can distinguish the process from the method$_n$ and thus (2) can distinguish process reliability from sensitivity. In addition, and very significantly in my view, these distinctions coincide neatly with the two kinds of knowledge-precluding epistemic luck: process luck and luck due to peculiar features of the actual world. (The emergent picture painted by my many thousand words appears, I trust, well motivated and coherent.) Grant, then, that White is a reliable oriole detector. In favorable conditions, her beliefs about the presence of orioles are true most of the time. Still, in any particular situation her belief may be insensitive. Suppose there are fake orioles sitting in the same tree as the real oriole on which White's gaze is focused, and that White cannot tell the fakes from the genuine orioles. If it were false that the bird-like thing White sees is an oriole, she would still believe that it is, based on her method$_n$: If the thing currently in my line of sight has such and so features, conclude that it is an oriole.

So long as processes abstract from the specific content of a particular perceptual belief, then, we can individuate them as narrowly as we like and still avoid the single case problem. What about non-perceptual, general beliefs, for instance that orioles have black and orange features? Since the specific content is itself general, it cannot be included in the process type. This is a welcome result, because it shows again that there is no danger of a conflation of processes and methods$_w$ with methods$_n$. The latter may be something like: If I read in the encyclopedia that orioles have black and orange features, conclude that they do. The process might be *forming beliefs about animals (birds? orioles?) on the basis of reading about them in the encyclopedia.*

It is clear, then, that an adequate response to the generality problem demands that process types be individuated specific-content-neutrally. We have seen that this makes reliabilism powerless to solve the Mary problem, which in turn provides a partial confirmation of our theory's inclusion of sensitivity. Before concluding this chapter, I want to apply the theory, with its criterion for individuating types, to one of the most difficult cases that we have encountered.

Haveit wants me to believe truly that someone in my office owns a Ford, so he tells me that Nogot has one when in fact it is only Haveit himself who owns a Ford. Had Haveit not owned a Ford, he would not have led me to believe that Nogot has one. Here again the belief satisfies sensitivity but does not constitute knowledge. Shope's suggestion, that the process used in believing what Haveit says could be characterized "as forming a belief guided by evidence that has been unsuspectedly fabricated," is compatible with our narrow, content-neutral criterion of process individuation, and that process is unreliable.[34] The problem, though, is that the evidence *having been unsuspectedly fabricated* is not causally operative in the process by which I form the belief that someone owns a Ford. What is causally operative, however, is that the evidence is provided by Haveit (or *that* guy, if I do not know Haveit). The process, or method$_w$, by which I form my belief is (or includes) *forming beliefs on the basis of evidence provided by Haveit.* This is neutral as regards the specific content of my belief. Though Haveit wants me to form one particular true belief, he is in fact lying to me, and thus his testimony is unreliable – on it I form beliefs some of which are false, for example, that Haveit owns a Ford. Once again we achieve the desired result, in this case not through sensitivity involving reference to methods$_n$, but through the narrowly individuated but content-neutral methods$_w$.[35]

C. Conclusion

My main objective here is to set out the theory that knowledge is reliably formed sensitive true belief and to defend it against actual and potential criticisms. Rather than rehearsing the main claims of the chapter, I should like to highlight the more general and, in my view, more interesting idea that reliability and sensitivity are natural complements. First, taken separately, they are each motivated by the fundamental notion that knowing requires distinguishing what is the case from what is not, or one thing from another. I argued in Chapter 4 that safety-based accounts of knowledge sometimes miss this point. Second, reliability traffics at a fairly high level of generality, and thus needs supplementation by a principle that captures the demand to account for the way in which knowledge-constituting beliefs are those that would not be held if the specific contents were false. Sensitivity, in its modal abstractness and its specificity, needs supplementation by a more general condition that appeals to our intuition that knowledge derives from general cognitive processes. Beliefs that satisfy sensitivity do not just pop up from anywhere or nowhere, and if they did, they would not be cases of knowledge.

Though it was not a central topic of this chapter, I also argued anew that there are resources in process reliabilism, properly understood, to answer objections brought against it by newer, perhaps more popular strains of reliabilism, such as agent-centered accounts and virtue theory. I have tried to undermine the view that knowledge entails an agent's creditworthiness while claiming that there are other means within process reliabilism to solve the serendipitous reliable process problem. That still leaves us with the value problem, which is the focus of the second part of the next and final chapter, but before that I will explain how the reliably formed sensitive true belief theory can take the counterintuitive sting out of closure failure.

6 Closure and the value problem

Having set out the theory that knowledge is reliably formed sensitive true belief in some detail in Chapter 5, I turn, finally, to considering two influential criticisms of this kind of view. First, many philosophers believe that any theory that entails the rejection of the principle that knowledge is closed under known entailment must be seriously flawed, given that the principle appears beyond reproach. The sensitivity condition implies the rejection of closure. However, though the closure principle does indeed seem true, I do not feel compelled to answer objections based solely on its strong prima facie plausibility, because the theory presented here responds attentively to our considered judgments about particular cases of knowledge. If the theory is otherwise correct and entails non-closure, then I am willing to live with that. But it is sometimes argued that the denial of closure has other unwanted consequences, and such arguments go beyond merely insisting that closure is true. In particular, I take up a challenge put forth by some contextualists: Because skeptical arguments often hinge on the principle of closure, denying that principle requires explaining how the skeptical arguments have seemed so powerful. That is, it is not enough simply to put forth a theory with anti-skeptical results one of whose consequences is non-closure; one must also give a plausible explanation of how closure *could* fail and why the skeptical scenarios introduced in skeptical arguments have seemed so bothersome, else one's position fails to take seriously the prima facie and historically vexing strength of those arguments. In §A, I aim to show that the theory that knowledge is reliably formed sensitive true belief not only entails non-closure but (1) *explains* how closure can fail, and (2) explains the intuitive appeal of skeptical hypotheses underlying the skeptic's arguments.

Second, relatively recently reliabilism has been linked to consequentialism in ethics, in that on both views what matters is the end achieved. Reliabilists think of the conditions other than truth and belief that constitute knowledge in truth-conducivist terms, which suggests that the only epistemic end (or goal) is truth. If that is right, then the complaint against reliabilism is that it cannot explain how reliably formed true belief, that is, *knowledge*, is any better to have than any other true belief, such as one based on mere guessing, since in both cases the end is achieved. But surely, it is claimed, knowledge is better than mere true belief. This is the value problem for reliabilism. Building on work by Wayne Riggs,[1] I argue in §B that shifting focus from the concept of reliability to the notion of an enviably strong epistemic position can help. As I use the terms, (possessing a) "reliable process" is roughly equivalent in reference to (being in a) "strong epistemic position," but thinking about a strong epistemic position rather than reliability allows us to emphasize the value of an agent being well situated to attain true beliefs. I will argue that we can recognize this value even when the agent fails to achieve true belief, hence that reliabilists can locate a source of value independent of truth. This is the general thrust of §B.

Before proceeding to the problem about closure, I want to enter into the record an important disclaimer. I am providing this chapter solely as a courtesy to the reader who finds the theory articulated in Chapter 5 promising and wants to hear more about how it might speak to these influential objections. This is important for two related reasons. First, if one thinks that the theory defended here answers adequately to our considered judgments about cases of knowledge, then it is at least *not unreasonable* to think it permissible to ignore the closure and value problems. If it is true that I can know that I am sitting here typing but not that I am not a BIV, even though I know the former entails the latter, then the closure principle is false. (I think it is important to emphasize again that almost all discussions of the closure problem begin with these claims about our knowledge statuses as more or less obvious, continue with some hand wringing about the implications for closure, and then usually try to alleviate the felt concern by finding a way to maintain closure through *overturning* one of the obvious claims from which the closure problem originates.) And why insist that any particular case of knowledge is more valuable than any particular merely true belief? (If truth is our aim, there are better and worse ways to get it, but once true belief is attained, it is valuable in itself and for whatever one might do with it. My father used to say, "Better lucky than good." I don't think he really meant it, but it's easy to see his point.) Second, especially with regard to the value problem, it follows that if one does not accept my responses to the problems discussed here – does not believe I have identified the right *kind* of value – and believes there are no acceptable answers from the standpoint of the theory, this is not enough to show that the theory itself is false. Instead, I think it reveals only a clash of views about the nature of the value inherent in knowledge. For example, an internalist who thinks that knowledge is valuable only because the agent fulfills her epistemic duty might deny that the kind of value I will describe in §B is adequate, and might therefore claim that reliably formed sensitive true belief is not knowledge. That goes too far. The account of value in reliably formed belief that I set forth below neither depends upon epistemic duty fulfillment nor entails that a particular case of knowledge is more valuable than its corresponding merely true belief, but it shows clearly how being in a position to know is better than attaining a true belief from a lesser position.

A. Closure

The theory that I have presented here is a kind of invariantism, one where the truth conditions of knowledge attributions do not vary according to features of the attributor's context. (Of course, whether one knows does depend on other features of context, as the barn facsimile case clearly shows.) In addition, it achieves an anti-skeptical result, showing how knowledge is possible, even if one cannot rule out radical skeptical hypotheses. Sensitivity plays a crucial role in this anti-skeptical strategy. Contextualists often claim the following two advantages over invariantist theories with this kind of response to the skeptic. First, the contextualist can explain why the skeptic's premises in the argument from ignorance, introduced in Chapter 4, §§C and D.2 and discussed further, below, are so convincing, hence how skepticism itself can seem so persuasive, whereas the theory that I have defended, particularly because of its sensitivity condition, *discounts* the intuitive appeal of the skeptic's argument by denying the principle that knowledge is closed under known entailment, which underlies a crucial skeptical premise in the argument from ignorance. Because the principle of closure appears to be unassailable, contextualists (and others) accuse the sensitivity theorist of (excessive) epistemological revisionism. Second, the contextualist claims that the truth conditions of knowledge attributions are context-dependent, and that this fact emerges from ubiquitous evidence of how knowledge is actually attributed. Non-contextualist theories fail to account for this evidence.[2] I shall formulate the argument from ignorance and the contextualist response to it (§A.1), present the reliabilism-based explanation of closure-failure (§A.2), and then briefly take up the offensive against contextualism (§A.3).

1. The argument from ignorance and the contextualist response

Let's begin with a formulation of the skeptic's argument from ignorance. "SH" stands for a skeptical hypothesis, for instance that I am a BIV being "fed" my experiences electrochemically (or whatever) – the ones I am having now, for example as of typing at my computer. "O" stands for any ordinary empirically believed proposition, for example, that I am typing at my computer.

Argument from ignorance (generalized version)

(P1) I don't know that not-SH.
(P2) If I don't know that not-SH, then I don't know that O.
(C) Therefore, I don't know that O.

P2 seems plausible because if my experiences do not suffice to rule out the possibility that I am a BIV, then they cannot uniquely "rule in" the fact that I'm typing at my computer. In the present context, however, P2 is question-begging, since the sensitivity theorist claims that I can know that O without knowing that not-SH. In nearby worlds where it is false that I am typing at my computer, I do not believe it (based on my actual belief-forming method). Provided also that my belief is reliably formed, I know that I am typing at my computer. Suppose, then, that I know that I am typing at my computer, and that I know that this entails that I am not a BIV. Do I know that I am not a BIV? If it were false, I would believe it anyway. Thus according to sensitivity, I do not know that I am not a BIV.

The argument from ignorance leads either to skepticism or, seemingly, to paradox, since it would appear that at least P1 or P2 must be false if we have any empirical knowledge, and both of the premises appear true. One way to deny the soundness of the argument from ignorance, a way adopted by the sensitivity theorist, is to deny P2. So the obvious question is: What grounds are there for accepting P2? The question brings us to an application of the principle of closure:

Principle of closure

(1) I know that O.
(2) I know that if O, then not-SH.
(3) Therefore, I know that not-SH.

(More generally: For any propositions *p* and *q*, if I know that *p*, and I know that *p* entails *q*, then I (am in a position to)[3] know that *q*.) If I know that I am sitting in front of my computer typing, and I know that this entails that I am not a BIV, then I (am in a position to) know I am not a BIV. Thus the principle of closure, if true (or (informally) valid, depending on whether one takes it as a sentence or syllogism) yields the following conditional (assuming that (2) is unassailable, which it is): If I know that O, then I know that not-SH. The contrapositive of this is: If it is not the case that I know that not-SH, then it is not the case that I know that O. Thus the principle of closure yields P2.[4] If I don't know that I'm not a BIV, then it seems I don't know that I'm typing at my computer. (Though the principle of closure is not formally valid, one can see how it gives rise to the debate between Moore and the skeptic – one person's (application of) *modus ponens* is another's *modus tollens*.)

Sensitivity theorists reject closure. Because many of my beliefs with respect to O-type propositions are sensitive, whereas those regarding not-SH-type propositions are not, I can know the former without knowing the latter. This is where accusations of excessive revisionism arise. The sensitivity theorist offers anti-skeptical results – I know lots of ordinary empirical truths – but denies the compelling principle of closure. Contextualists derive from this supposedly unmotivated denial a criterion of adequacy on any theory of knowledge, of which each of the following, from Stewart Cohen and Keith DeRose, respectively, is an expression:

> In some sense skepticism is crazy. So we want a response to the paradox that preserves our belief that we know things. But if such a response is to constitute a resolution of the paradox, it must explain the undeniable appeal of skeptical arguments.[5]

"We must explain how two premises [P1 and P2] that together yield a conclusion that we find so incredible [C] can themselves seem so plausible."[6] DeRose and Ram Neta go even further, claiming that we must also explain how some skeptical hypotheses are more compelling than others.

> Hypotheses are supposed to explain; skeptical hypotheses should explain how we might come to believe something despite its being false. ... [W]hen we encounter effective skeptical hypotheses, we have some understanding of how (if [SH] is true) we have come falsely to believe that O.
>
> (DeRose 1999, 197)

Neta explicitly lays out the following among the criteria of adequacy satisfied by his account in solving the skeptical puzzle:

> It explains why [I do not know that I am not a BIV] is more plausible than [I do not know that I have hands], and thereby explains why the BIV hypothesis can be used to generate the skeptical puzzle but the mere "I don't have hands" hypothesis can't be.[7]

There seem to be a number of related objections here to any view that rejects closure. First, (P2) seem plausible, but as we have seen, it hinges on closure, which seems plausible to the same extent. So any view that denies closure must not only entail that denial, but also *explain* its possibility. Second, any view that says I can know that O while not knowing that not-SH seems, at least *prima facie*, to ignore the intuitive force of the skeptical argument. We do not merely want to brush aside skeptical hypotheses while claiming ordinary knowledge, which would be to deny closure "on the cheap"; rather, we need to explain the worrisome nature of those hypotheses and how they are nonetheless compatible with knowledge. Relatedly, we need to explain *how*, if such an hypothesis were true, knowledge of ordinary propositions would be undermined. (See also note 15.) Finally, in order to answer the second problem, it would be nice to have a means of explaining the difference in intuitive force between compelling skeptical hypotheses (for example, that I am a BIV) and less forceful ones (for instance, that I do not have hands).

Before defending the reliably formed sensitive true belief theory against the charge that it cannot accomplish all this, it would be well to see briefly how contextualism answers the skeptic while satisfying this close-knit family of criteria. The contextualist claims that the truth of knowledge attributions depends in some way on the attributor's context.[8] In most ordinary contexts, where skeptical hypotheses are not under consideration, it is true that S knows that O provided that O is true, S believes that O, and S is justified or warranted in believing that O. When skeptical scenarios are introduced, perhaps merely by being suggested, the context changes and the standards for true knowledge attributions are raised. Because one cannot rule out skeptical hypotheses, one does not know, in these contexts, the ordinary empirical truths that one knows in more mundane contexts. In both kinds of context, the principle of closure holds. Thus in ordinary contexts, I know both that O and that not-SH (so long as the possibility that SH is not entertained), but in skeptical contexts, I know neither O nor not-SH.

How does the contextualist satisfy her criterion of adequacy according to which it must be explained why some skeptical hypotheses are more compelling than others – why is, for example, "I do not know that I am not a BIV" better suited to generate the skeptical problem than "I do not know that I have hands"? Here I shall stick to DeRose's story, which requires setting out some of his machinery. The context of attribution determines the range of epistemically relevant worlds through which my belief must be sensitive if the knowledge attribution is true. Interestingly, then, DeRose includes a sensitivity component, but it is not a necessary condition for knowledge. What is necessary is that one be in a strong enough epistemic position with respect to the proposition in question: "The further away one can get from the actual world, while still having it be the case that one's belief matches the fact at worlds that far away and closer, the stronger a position one is in with respect to P" (DeRose 1999, 204). What, then, constitutes an epistemic position strong enough for knowledge? This depends on context, more specifically, on the rule of sensitivity: "When it is asserted that some subject S knows ... some proposition P, the standards for knowledge ... tend to be raised, if need be, to such a level as to require S's belief in ... P to be sensitive for it to count as knowledge" (DeRose 1999, 206). In ordinary contexts, ones where skepticism is not being considered, the standards for knowledge are somewhat low in that one's beliefs need to track the truth only throughout nearby worlds – far off skeptical possibilities are irrelevant. If the question concerns whether I know that I am typing at my computer, the answer is yes because, if it were false, I would not believe it. Because the required strength of epistemic position is determined by the rule of sensitivity, that standard applies to knowledge of any proposition in that context – the belief must match the fact throughout the set of worlds determined by the standard. Well, my belief that I am not a BIV is true throughout the worlds that are relevant in a non-skeptical context, and so I know it too. (This is true so long as it's not explicitly stated or thought, which ought to be a larger concern for the contextualist than it usually appears to be; I will take up this issue below in §A.3.) The contextualist thus upholds closure. In ordinary contexts, I know both that

O and that not-SH. Notice, too, that my belief that not-SH is not itself sensitive, for if it were false, I would believe it anyway. This demonstrates how, on DeRose's view, sensitivity is not a necessary condition for knowledge.

When the skeptical worry expressed by "I don't know that I am not a BIV" is raised, the range of epistemically relevant worlds is expanded from those operative in ordinary contexts. The rule of sensitivity kicks in, raising the standards of epistemic strength high enough to require sensitivity to the proposition that I am not a BIV. But my belief that I am not a BIV is insensitive – if it were false, I would still believe it. Thus when such a skeptical hypothesis is raised in conversation, I don't know that it is not true, and, given the principle of closure, in those contexts, I also don't know that I am typing at my computer.

We are now in position to see how the contextualist differentiates compelling skeptical hypotheses from unthreatening ones. Consider an alternative, less compelling skeptical hypothesis, that I do not have hands. This does not enlarge the range of epistemically relevant worlds beyond those operable in ordinary contexts. I am sensitive to such empirical falsities, and so the right reply to this kind of skepticism is to say, "I do know that I have hands. Look here." (This may be accompanied by some Moorean hand-waving.) This sort of skeptical worry doesn't get off the ground because raising it does not create a context in which the range of epistemically relevant worlds is extended beyond those to which my belief about the (weak-kneed) skeptical proposition is sensitive. (In the nearest possible worlds where I do not have hands (think: horrible accident), I believe that I do not have hands, and in the nearest worlds where I do have hands, I believe that I do.)

In addition, the contextualist explains the air of paradox by claiming that we do not always realize that the truth conditions of sentences involving the expression "know" are context sensitive. The meaning of "know" in the sentence "I know that I am typing at my computer" is (usually) different than that in "I don't know that I'm a BIV" because the contexts of these self-attributions are normally very different. Because it takes an able and novel contextualism to straighten out this confusion, it is unsurprising that the skeptical argument has had such a grip on us.

2. Reliabilism, sensitivity, and the skeptical premises

What has the sensitivity theorist to say about the alleged criteria of adequacy set out above in §1? That is, why are the skeptical premises so plausible? Before we can answer that question, we need an account of closure failure – ultimately, of the denial of the second premise of the argument from ignorance, which will take some patience to build. Sensitivity plays a crucial role here, but the lingering worry is that if closure failure hinges on sensitivity, then sensitivity must not be a necessary condition for knowledge because denying closure commits one to excessive revisionism. I shall argue that, in cases of closure failure, sensitivity is violated for beliefs in the denials of skeptical hypotheses because there is no reliable cognitive process upon which we base such beliefs. My aim is to show that sensitivity is not a stand-alone principle to be jettisoned if ever it leads to an unintuitive consequence, but is rather an integral part of a reliabilist epistemology.

The explanation of closure failure is our first objective, but it is important to see that the process reliabilist component of our theory, by itself, cannot provide this. A belief-forming process is reliable if it produces mostly true beliefs throughout nearby worlds. If I have a reliably formed true belief that I am typing at my computer, and I infer that I am not a BIV, then it would seem that my belief that I am not a BIV is reliably formed because it too is true in nearby worlds. Reliability by itself does not deliver non-closure.

One might think it reasonable at this point to accept a straight reliabilist account of knowledge and reject the sensitivity condition, since we've achieved a very strong anti-skeptical result, one that yields knowledge of even anti-skeptical propositions, and we've maintained closure. But this violates one of the contextualist's criteria of adequacy in answering the paradox based on the argument from ignorance because it fails to take seriously the intuitively plausible claim that we do not know the denials of skeptical hypotheses. After all, all of my evidence that I am typing is compatible with my being a BIV, so how can I infer from that evidence that I am not? And if I cannot do that, then how does my evidence sufficiently warrant my belief that I am typing at my computer? (For all I know, I'm a BIV!)[9] Besides, sensitivity is the crucial anti-veritic luck component of our theory.

The first step in accounting for closure failure is an argument that the knowledge-constituting warrant for my belief that I am typing – for ordinary propositions generally – is not sufficient for, and therefore does not transmit to, my belief that I am not a BIV. (Warrant does not transmit in general to anti-skeptical propositions.) Crispin Wright and Martin Davies have both offered stories about failure of warrant transmission from premises to known entailments.[10] According to Wright, warrant transmission succeeds when one has independent warrant for the premises of an argument, and *thereby* has a warrant to accept the conclusion. Warrant does not transmit when the independent warrant for a premise is not sufficient for knowledge, and what would be sufficient is a prior, additional warrant for accepting the conclusion. The conclusion-independent warrant for my belief that I am typing is that I see and feel that I am. But this is insufficient for knowledge unless I am also entitled to believe that I am not a BIV (where "entitlement" is a very basic kind of warrant if I understand Wright correctly). But if I am entitled to that, and much of what Wright says suggests that I may be, then I cannot use my knowledge that I am typing to *gain* knowledge that I am not a BIV, for the latter is presupposed in the former. Any such inference would be question-begging.

On Wright's account, however, failure of warrant transmission presupposes closure, rather than undermines it. Warrant fails to transmit precisely in cases where one is entitled, hence presumably warranted, in believing the denials of skeptical hypotheses, from which it follows that, when I know that I am typing, I also know that I am not a BIV. But the view I have been advocating implies a rejection of Wright's claim that my warrant for believing that I am typing is insufficient for knowledge without an independent entitlement to the belief that I am not a BIV, while still capturing the insight that whether I have sufficient warrant for my ordinary beliefs hinges on its *being true* that I am not a BIV.

Jessica Brown has argued this very point. She claims that the epistemological externalist need not accept Wright's view that one is warranted in believing ordinary propositions only if one has a prior and independent entitlement to believe the denials of skeptical hypotheses.[11] What constitutes warrant for the externalist is that in fact one's belief is appropriately tied to the truth, such that if my belief that I am typing is reliably formed – that is, formed by a process that produces mostly true beliefs throughout nearby worlds – it is warranted. (Of course, on my view, it must be sensitive, too.) Externalist warrant for believing that I am typing thus requires that I am in fact not in a BIV world, but does not require that I am entitled to believe this. On this view, there is sufficient warrant for believing ordinary propositions independent of further warrant for believing the denials of skeptical propositions.

It follows that my warrant for believing ordinary propositions may not suffice to warrant my belief that I am not a BIV, and this would be an instance of failure of warrant transmission across known entailments. (Again, this is not the same type of transmission failure that Wright describes, where full warrant for a premise hinges on an antecedent warrant for the conclusion.) That it is a fact that I am not a BIV is a necessary condition for my warrant to believe that I am typing, but it certainly does not add up to a warrant for believing that I am not a BIV. Wright thinks this kind of position is unintelligible, but his complaint is based on lack of appreciation for a properly externalist epistemology:

> I have a particular kind of warrant for believing *P* [for example, that I am typing] but, for all I am entitled to suppose, it may be that conditions are unsuitable for getting a warrant for *P* of that kind [conditions where, for example, I'm a BIV]. And that cannot be a clear-headed claim. If I assent to it, what can I possibly imagine entitles me to claim the warrant for *P* in the first place?
>
> (Wright 2003, 61)

The externalist may be perfectly happy to say that it would sound odd for me to *claim* that I have a warrant for believing an ordinary proposition when I am not entitled to believe I am not a BIV. Willingness to make claims typically involves adducing grounds or reasons to which one has access, so unless I take myself to be entitled to believe I am not a BIV, I could wonder whether my belief that I am typing at my computer is (fully) warranted. But the externalist ought nonetheless to say that my belief that I am typing at my computer is in fact warranted – reliably formed and sensitive – even if there are necessary conditions for warrant that are opaque to me.

We now have the required elements for closure failure. My belief that O is reliably formed and a case of knowledge. My inference to not-SH produces a reliably formed belief, in that the process produces mostly true beliefs throughout nearby worlds, but the inference does not provide knowledge, because the warrant I have for O (which requires that not-SH) insufficiently warrants my belief that not-SH. The sensitivity component of our theory somehow predicts this result – I do not know that not-SH because, if it were false, I would believe it anyway. But are we not back to square one? We are diagnosing closure failure by the application of sensitivity to O and not-SH, and the worry is that this is a better reason for rejecting sensitivity than for denying closure.[12]

In fact, we are not at square one, because we can *explain* why sensitivity is violated for not-SH type propositions by reference to the very general notion of reliability. Though my belief that not-SH is reliably formed when inferred from O, it does not constitute knowledge because its warrant is not transmitted from my belief that O. Why not? If we ask whether I know that not-SH, we are bringing into question a possibility, that SH, which, as Wright says, "is incompatible with the reliable operation of the cognitive capacities involved in generating the warrant" (Wright 2003, 62) for my belief that O. If the possibility that SH does not obtain, then my environment is compatible with the reliability of my cognitive capacities, but it is because I have no reliable capacity for distinguishing SH worlds from not-SH ones that I lack knowledge that not-SH. For good reasons, reliability is characterized in terms of truth-conduciveness in nearby worlds. This gives the superficial impression that my belief that not-SH is reliably formed because it is true throughout nearby worlds. However, the only plausible warrant for this belief that does not dismiss skepticism out of hand by claiming entitlements to know not-SH is through inference from my belief that O. We have seen that this does not always work, because the warrant for O, which requires that I am in fact not a BIV but not that I am entitled to this belief, does not transmit to a warrant for believing I am not a BIV. One could go along with Wright and claim that I have warrant for believing that I am not a BIV, perhaps in the form of an a priori entitlement, but I submit that this really does fail to give the skeptic her due. If it were true, it would be difficult to see why skepticism was ever a vexing position.

Closure fails because the independently sufficient warrant for believing that O does not transmit a sufficient warrant for believing not-SH, even though both beliefs are reliably formed according to our definition of reliability. So the situation appears to be that, while my belief that not-SH does not directly violate the reliability condition, there is *something* unreliable about the way I obtained it. If the sum and substance of the diagnosis of my lacking knowledge that not-SH were that the belief is insensitive, the reply might very well be: "So much the worse for sensitivity. It is merely an abstract principle that tells us little about how knowledge is achieved. If it's causing you trouble, for instance in committing you to denying closure, then reject it." I think we can block this response by showing that sensitivity failure is intimately related to unreliability, thereby wedding the two key features of the belief-truth link in my account, and thus motivating sensitivity as part of a package reliabilist deal. The core notion of both sensitivity and reliability is responsiveness to the world – the capacity to distinguish truth from falsity – and so we ought not to detach sensitivity if it yields prima facie untoward consequences. It is a vital part of an overall reliabilist epistemology.

Why did we define reliability as truth-conduciveness throughout *nearby* worlds, rather than in just the actual world or throughout all possible worlds? First, we rejected actual world reliabilism because we wanted to capture the intuitive idea that knowledge requires being able to distinguish *p* from possibilities where not-*p*. A reliable belief-forming process is not only one that delivers true beliefs in the actual world where *p*, but one that would produce true beliefs in nearby worlds, including the nearest not-*p* worlds. Second, reliabilism cannot require truth-conduciveness throughout *all* possible worlds, because worlds where radical skeptical hypotheses are true are worlds where our normally reliable cognitive processes are useless. Requiring truth-conduciveness throughout all possible worlds fails to take seriously the very idea of a reliable belief-forming process. Our cognitive processes are not truth-conducive in some distant worlds.

So the notion of a reliable belief-forming process entails truth-conduciveness throughout nearby worlds. This opens a kind of loophole in the reliabilist account. To exploit it, pick a proposition, for instance that I am not a BIV, that explicitly raises a counter-possibility that holds only in some distant world. Since I form this belief in the actual (hopefully non-BIV) world, and it is true throughout nearby worlds, it does not violate the reliability condition. But surely the notion of reliability was not intended to extend to such cases, for I cannot distinguish between BIV worlds and non-BIV worlds; my reliable belief-forming processes are not responsive to the difference between BIV and non-BIV worlds. Sensitivity closes the loophole by honoring the basic reliabilist insight that knowledge requires processes capable of distinguishing *p* from counter-possibilities where not-*p*. If I were a BIV, I would not believe I am, and that is why I don't know. In short, my belief violates sensitivity because it is not *in fact* reliably formed, though the truth-conducivist gloss on reliability itself does not make this explicit. Thus this is one more way in which sensitivity nicely complements reliability – it gives a fuller picture of the nature of reliably formed belief. It is thus motivated not solely by the results it achieves in particular cases, but by its place in the larger reliabilist project.

Pritchard does not see things this way (Pritchard 2005, §2.3). Instead, he criticizes sensitivity theorists for an asymmetry in their application of intuitions about which alternatives are relevant. If the possibility that I am a BIV is irrelevant, because too distant, to whether I know that I am typing at my computer, then it constitutes a double standard to say that it *is* relevant to whether I know I am not a BIV. But, as I see it, the core notion of reliability is not captured by application of a uniform set of relevant alternatives, but by responsiveness to the world – the capacity to distinguish. As such, it is not an appeal to double standards that allows us to think of different possibilities as relevant to knowledge of different propositions; we apply a univocal standard that requires distinguishing *p* from not-*p*, for that is what reliability consists in. The alternatives that are relevant, in every case, include the closest worlds where not-*p*. And as against contextualism, they are not determined by features of the attributor's conversational context, but by the nature of the proposition believed to be true.

The case against closure, one that motivates the sensitivity-based denial of closure within a reliabilist framework, is now complete. We are thus prepared to assess the contextualist's criticisms regarding the inability of a theory that denies closure to explain the plausibility of the skeptical premises in the argument from ignorance, which include the claims both that one does not know the denials of skeptical hypotheses and that, by application of closure, this threatens knowledge of ordinary truths.

Why, then, does closure seem so plausible? First, and most obvious, the propositions embedded in the "knows that" operator constitute a valid argument. So it would seem that if one knows that the propositions constituting the premises are true, one can just deduce the conclusion and thereby know it. In fact, this is almost always the case. It won't be the case when one of the *necessary conditions* for knowing an ordinary proposition O is that some proposition, such as not-SH, is true, but the agent has no way of telling whether not-SH is true, and yet infers it from O. Externalism is thus uniquely suited to deny closure. One need not have a reason to believe (or be warranted in believing) that not-SH is true; provided that it is, one can know that O if one is independently warranted in believing it.[13]

So why is it that, even if we accept or at least entertain externalism, we still find the second premise of the argument from ignorance compelling? I think part of the reason is that the version of closure that I presented, whose form is dictated by the need to fit the argument from ignorance, which is essentially taken from DeRose (1999), is stated in the first person. It is very easy to look at the claim, "I know that O," and take myself to know it – to know that I know that O. (After all, this is what we are doing when assessing, from the theoretical perspective, the closure principle.) And it is a short step from there to think that, if I know that I know it, I must know that the conditions possible for knowing, for instance that not-SH, are fulfilled. I add the known premise that if O, then not-SH, and I deduce the conclusion, feeling no compulsion to question my conviction that not-SH. My suggestion is that, stated in the first person, the closure principle and the argument from ignorance propel us to think along *internalist* lines, and it's difficult to kick the habit.

Suppose the closure principle were stated like this:

S knows that O.
S knows that O entails not-SH.
Therefore, S knows that not-SH.

Now when I, or you, reader, consider closure in this form, its allure is not *as* powerful, at least insofar as the compulsion to think internalistically is diminished. It is at least somewhat easier to pause and say, "Well, maybe S's warrant for believing O does not include any independent warrant to believe not-SH. If not, she will not be inclined to think she can merely deduce not-SH, because, as Wright says, it may sound odd for her to assert that she knows that O in the first place, even though she does know it." From a "god's-eye" perspective, there are cases where it will be true to say that S knows that O, but from S's own perspective, she may be more reticent.

My conjecture is that internalism has held such sway in epistemology that, even when we are prepared to reject it, we lapse into thinking internalistically. This will be especially so when we put ourselves in the shoes of the agent making claims to knowledge, since it is all too easy then to take ourselves to know that the necessary conditions for knowledge are fulfilled. When we do this while considering arguments embedded in "knows that" operators that we know independently to be valid, it is extremely difficult to think that the conclusion – namely, that we know the known consequences of known premises – does not follow automatically. So *of course* closure seems plausible.[14]

Our next task is to explain why skeptical hypotheses are so compelling. This is a slam-dunk if one accepts my view. They are so compelling because we do not know that they are false – if they were true, we would still believe they are false. And we can also explain why some skeptical hypotheses are bothersome while others, such as the hypothesis that I do not now have hands, are not. Sensitivity handles these cases nicely. If it were false that I have hands, I would not believe that I do. I would see and feel that I do not. So these simple skeptical hypotheses do not worry us at all.

Recall too, from §A.1, that contextualists demand an explanation of how, if skeptical hypotheses were true, we would come to have false beliefs.[15] We have already briefly touched on this. If I were a BIV, I would have lots of false beliefs because BIV worlds are incompatible, just as Wright says, with the reliable operation of my cognitive capacities. And this, too, is part of the explanation of the difference between powerful and lame skeptical hypotheses. I do have a reliable means of forming beliefs about whether I have hands, and so the hypothesis that I do not does not move me at all. But I know that if I were in a vat world, my otherwise reliable belief-forming processes would be undermined, issuing in widespread false beliefs. But this isn't so for just any skeptical hypothesis. It turns out that those that are not epistemologically bothersome are the ones that would not, if true, issue in false beliefs.

Good skeptical hypotheses – really bothersome ones! – appeal to scenarios where the normally reliable cognitive apparatus I *would* use to form beliefs is useless. No wonder this causes angst! Thus the BIV hypothesis seems more compelling – witness its pervasive use in these discussions – than the demon hypothesis. It is rather easy (for philosophers anyway) to imagine someone using my normal cognitive capabilities against me, having knowledge of their workings. (Put my brain in a vat; set up some fake barns or paint some mules.) DeRose would be right to say that the demon hypothesis is unlike the hypothesis that I don't have hands with regard to the sensitivity of my respective beliefs. But the demon hypothesis is also unlike, but in a different way, the BIV hypothesis. The demon hypothesis seems unmotivated because we are not told how the demon turns the trick. (The demon hypothesis offers little more than the simply stated worry that, for some reason, I might be wrong about everything.) What is the mechanism by which the demon furnishes our experience? The BIV hypothesis adds something, albeit (hopefully) fictional, to the story. My very brain, the seat of my normally reliable cognitive powers, is being used "against" me such that its normal functioning is rendered useless to settle the burning question. I take it that this is why philosophical commentary about the film *The Matrix* has become its own industry – there really is something (almost?) compelling about the idea of our cognitive powers used against us to make us believe falsely. And of course, if this is right, it also explains *how*, as DeRose demands, if a skeptical hypothesis were true, ordinary knowledge would be undermined. (The explanation goes beyond citing the obvious fact that my beliefs would be false. It explains *how* I go wrong.[16])

Before concluding this section and moving to a brief and very general critique of contextualism, I want to highlight a basic claim made by both DeRose and John Hawthorne, each of whom seeks to maintain closure. Hawthorne's recent book, *Knowledge and Lotteries*, is especially good at showing how any theory of knowledge has counterintuitive consequences. Suppose I just got home from grocery shopping and parked my car in the driveway. It is plausible to say that I know that my car is in the driveway. But I also know that this entails that the car has not been stolen since I got home, and yet it is also plausible to say that I do not know that it has not been stolen. (I haven't checked in the past few minutes.) Similarly, unless we accept full-blown skepticism, it seems I can know that I'm typing, that this entails that I'm not a BIV, but I do not know that I'm not a BIV. Here are our options. (1) Maintain closure and accept that I do not know my car has not been stolen and that I do not know I'm not a BIV, and thereby accept widespread skepticism, according to which I also do not know that my car is in the driveway and that I am typing at my computer. (2) Maintain closure while denying skepticism, which yields knowledge (through known entailment) that my car has not been stolen and that I am not a BIV. But I don't know these things! (3) Deny closure, making possible knowledge of ordinary propositions but not of anti-skeptical ones. But denying closure seems intuitively wrong, so something has to give. Similarly, the springboard for DeRose's contextualism is the argument from ignorance, and he claims that the premises are utterly plausible while the conclusion, skepticism about even ordinary propositions, goes too far.

My point in raising anew the general dialectical structure of the argument from ignorance is simply to show that even those who refuse to deny closure accept that our claims to ordinary knowledge are very strong and any claims to know the denials of skeptical hypotheses seem entirely unfounded. Both DeRose and Hawthorne agree that "something has to give," and both offer good reasons for not denying closure. Even so, the intuitive data about our knowledge claims very strongly supports an inquiry into the possibility of closure failure.

3. A clash of intuitions

The tenor of this chapter so far has been somewhat defensive. Now I want to briefly take up a very general line of offense against the contextualist. What indeed are the criteria of adequacy in solving the skeptical puzzle? Contextualists offer theirs, among which is adhering to closure, but here it seems one's intuitions are required to do a lot of work. That's fine, but I would at least like to point out some of the contextualist's claims that do not answer to everyone's intuitions.

According to the contextualist, it may be correct to say both that S knows that *p* and that S does not know that *p*, though S has exactly the same evidence, and all the external facts except conversational context, or perhaps the aims and purposes of the knowledge attributor, are the same in both cases. This result is explicit in an example from Cohen.[17] Suppose that Mary and John are at LAX waiting for their flight to New York (LAX). They want to know whether the flight stops in Chicago, and they overhear a conversation in which another passenger, Smith, says he knows, on the basis of the flight itinerary he received from his travel agent, that it does. Does Smith in fact know that the flight stops in Chicago? Suppose that Mary urgently needs to stop in Chicago because she needs to pick up a kidney on the way to New York, to be transplanted in a family member. (Here I deviate, without consequence, from the original example.) She may then decline to attribute knowledge to Smith, citing the possibility that the itinerary has changed. Smith's conversational participants, however, may truthfully attribute knowledge to Smith because they have no pressing reason to stop in Chicago, and thus no reason to mistrust the itinerary. In sum, on Cohen's view: "When Smith says, 'I know … ', what he says is true given the weaker standard operating in that context. When Mary and John say 'Smith does not know … ', what they say is true given the stricter standard operating in their context" (Cohen 2000, 97). The only difference between the two contexts is that Mary's reason for wanting to know is more pressing because she needs to pick up the kidney, and so her standard for knowledge is higher.

An important question arises: Why is the following not an abominable conjunction? Smith knows that the flight stops in Chicago, and Smith does not know that it stops in Chicago, on the very same evidence and with the very same facts (excluding conversational context and agents' purposes) in place. Now, I am not denying that we often *talk* this way, but rather than accepting this as data upon which to build a theory of knowledge, perhaps we should question our ordinary knowledge attributions, since they can sometimes yield contradictory verdicts as to a thinker's knowledge status, even where her evidence and the surrounding facts are the same.[18] Of course, the contextualist has a reply to the apparent contradiction involving Smith's knowledge status, which amounts to saying that "knows" means something different in different contexts, and so there is no real contradiction. In order to avoid what strikes me as an "abominable conjunction," the contextualist goes in for her own epistemological revisionism: (1) where the skeptic's challenge arises only in skeptical contexts, which misses the point of skepticism – it aims to undermine knowledge in ordinary contexts[19]; (2) where normal speakers sometimes fail to know what they mean by "know" because they can fail to see the context-sensitivity of knowledge attributions[20]; and (3) where it can be true at one and the same time, for example, that Smith both knows and does not know that the flight stops in Chicago. While it may be admitted that denying closure defies our intuitions, the contextualist also strains epistemological intuitions. Is contextualism worth the price?

DeRose and other contextualists would answer "yes," citing evidence of how knowledge attributions are commonly made. Such evidence shows that we in fact attribute knowledge in just the context-sensitive ways predicted by contextualism (even if we do not realize that we are equivocating on "knows"). But then again, we often attribute knowledge to animals; for example, the cat knows the squirrel is in the tree. A reliabilist can account for this, provided that there is a reliable mechanism that produces a "squirrel-in-the-tree" representation[21] in the cat. Given the plausible claim that cats don't think about skeptical scenarios, it is difficult to see what contextualism would add to our understanding of the cat's knowledge.[22] Now, perhaps we are wrong to attribute knowledge to cats, but that is in fact part of the point I am trying to make. It is dangerous to build a theory around ordinary linguistic phenomena because the ordinary *uses* of the term "knows" one should use as intuitively correct data for the theory are contestable. This is consistent with claiming that our best (most stable and shared) intuitions can and should guide us in theory construction.

In any case, it is not obvious that reliabilism fairs worse than contextualism in accommodating ordinary linguistic practice. People attribute to themselves and others knowledge of lots of empirical propositions while denying knowledge of skeptical hypotheses – at least those who have seen *The Matrix* or patiently listened to a philosopher describing a BIV scenario. Here the reliabilist and contextualist agree. Contextualists then appeal to "data" from cases like Cohen's story about Mary's flight from LAX to show that knowledge attributions are context-sensitive. But what are the relevant data here? Perhaps we should ask: Would reflective adults agree that those knowledge attributions, dependent as they seem to be on context, are true? It is not implausible to think that many would say "no" simply because there is no difference in warrant or justification (or evidence or grounds) in cases where only conversational context shifts or agents' purposes are different. It is an open question whether epistemological theory should answer to folk knowledge attributions, and another whether the contextualist has an accurate reading of the data in the first place.[23]

Finally, contextualists claim supremacy in solving the skeptical puzzle by explaining the intuitive *appeal* of P1 and P2 while sometimes being able to deny C. Thus it is important for them to honor P2 and the closure principle that grounds it. In ordinary contexts, then, P1 (I don't know that not-SH) is false and C (I don't know that O) can then be false. In skeptical contexts, P1 and C are true. But recall that the skeptic's argument, according to the contextualist, is very persuasive. To defeat it, one of the two very plausible premises has to go. Now, my "intuition" is that there simply is no context in which one knows that one is not a BIV. How could I know *that*? The contextualist, on the other hand, concludes that sometimes I do know that I'm not a BIV – only at those times where I'm not thinking about it and no one is talking about it. But does this not do violence to P1, just as the contextualist claims that the reliabilist unwarrantedly denies P2? A clash of intuitions obviously cannot settle this matter, but settling it is not my aim. Rather, treat this as a kind of *tu quoque* argument against the contextualist.

The contextualist will reply that in philosophical contexts, where the stakes are raised, I do not know that I am not a BIV, which is how the skeptic's premise is respected. Fair enough, I suppose, but in maintaining the closure principle and a general anti-skeptical strategy, the contextualist claims that I sometimes know that I am not a BIV. I have no earthly idea how I could know this (provided the case is described sufficiently well).[24] Here, it seems to me, it is the contextualist who short-changes the skeptic, and so rejecting closure seems at least as plausible as claiming that I do know that I am not a BIV. In any case, assuming contextualism, how exactly do I "know" the anti-skeptical claim? I can't say it. I can't think it explicitly. (If I do, my supposed knowledge vanishes.) What exactly do I know?[25] There is a tinge of irony here for the contextualist. The whole account is driven by the desire to accurately capture the truth conditions of knowledge attributions. It is then claimed that in ordinary contexts one can know that not-SH, so long as that knowledge is not attributed! At the end of the day, is it more plausible to claim that we can know something as long as we don't think or talk about it, or to deny the principle of closure? If you believe that reliabilism fails to satisfy the criteria of adequacy that contextualists claim for solving the skeptical problem, you may opt for the former. I have tried herein to answer the contextualist, thereby giving reasons for choosing the latter.

B. The value problem

The recent turn toward virtue epistemology is motivated, at least in part, by the demand to capture the added value of knowledge over mere true belief. Specifically, it is sometimes claimed that reliabilism is incapable of explaining the value of knowledge because it is essentially consequentialist. Reliabilism focuses solely on the truth-conduciveness of the belief-forming process as the relevant connection between belief and truth that constitutes knowledge. But if true belief is the sole epistemic good, then, it is asked, why is it more valuable to have a reliably formed true belief than just any old true belief, based on hunches or lucky guesses or wishful thinking? Either way, the end is achieved. Let us call this the value problem for reliabilism.[26]

For openers, it is not entirely clear that there is a problem here, since one plausible reply might be that a particular case of knowledge does not entail more value than its corresponding merely true belief. This is nonetheless compatible with the added value cited by responsibilist epistemologists that accrues to an agent when she does her epistemic duty, or otherwise has a truth-seeking epistemological motivation, in *some* cases of knowledge. Assuming, however, that there *is* a problem here, it encourages the turn to virtue theory.[27] The added value is not in the belief itself, but in the agent. We recognize the added value of knowledge when we see that the agent deserves *credit* for her true beliefs in virtue of some intellectual character trait, or virtue, of her own, from which she forms her beliefs. However, without further details about what constitutes a virtue, this view is sometimes thought to be compatible with a more sophisticated version of reliabilism that effectively appeals to character traits, according to which a reliable process is one that produces true beliefs not only in the actual world, but also in most nearby possible worlds. We have seen this move before, which is essentially that from actual world reliabilism to modalized reliabilism. Recall Plantinga's brain lesion case. An agent has a brain lesion that causes her to believe she has a brain lesion. We may suppose that this is just a fluke. If, in the actual world, the brain lesion causes a true belief, but in most nearby worlds it does not, then the agent's belief is only accidentally true. Knowledge requires that the reliable process is not this unstable – it must produce true beliefs in most nearby worlds as well. By modalizing the process in this way, one effectively achieves reference to a stable character trait.

Riggs (2002) argues that reliabilism thus does capture a dimension of value, in that reliably formed true beliefs are not merely lucky, and so indicate a creditworthy achievement of the believer. He goes on to express qualms about whether this adequately secures the value requisite for knowledge. Zagzebski argues that it does not and offers a more demanding standard for virtue. Not only must the intellectual character trait be stable, it must also be motivational. Knowledge is more valuable than reliably formed true belief because one has knowledge only when one has an intellectual virtue that motivates her to seek the truth. This is a robust notion of virtue akin to moral virtue, where an agent's action exemplifies moral virtue only if the action is deemed praiseworthy, and the latter is true only if the agent is properly motivated to so act.

The shift to virtue theory, then, whether construed along externalist, agent reliabilist lines as in Greco and Sosa, or in Zagzebski's motivational terms, is intended to capture some measure of an agent's creditworthiness, but there remains the issue about when credit is in fact due. Riggs argues that reliabilism does explain an aspect of creditworthiness insofar as the agent's reliably formed true beliefs are not merely lucky and are attributable to the agent possessing a reliable process. Zagzebski would claim, however, that credit has not (or may not have) entered the picture yet, for how can one earn credit for some good outcome if she is not motivated to achieve it – when, in forming a true belief, she is not also motivated to achieve true beliefs? No credit, no value; no value, no knowledge.[28]

While Riggs is not a reliabilist, he does reliabilism a service by showing that the added value of reliably formed true belief lies in its ability to eliminate epistemic luck, which has been an important concern throughout this book. Unlike guesses or hunches that turn out to be true, reliably formed beliefs are not merely accidentally true. Better still, processes that are modally reliable eliminate the kind of luck that accrues to the brain lesion case. In demonstrating this positive feature of reliably formed belief, Riggs changes the focus of reliabilism from the end, true belief, to the agent who possesses the process. This partially undermines the case against reliabilism, the case being motivated by the thought that, if the end is all that matters, then reliably formed true belief is no better than true belief *simpliciter*. But then we have to contend with Zagzebski's point, which seems to be that this is a distinction without a difference, because credit implies praiseworthiness, which implies that the agent must be motivated to achieve true beliefs.[29]

I think Riggs is right that, if reliabilism is able to explain the value of knowledge, it cannot appeal simply to true belief as an end, precisely because that strategy is powerless to distinguish one kind of true belief from another. I think too that Riggs is right to focus on the anti-luck feature of reliabilism as a source of epistemic value. However, I do not want to follow

Riggs in his *diagnosis* of that value when he claims that it derives from credit due to the agent, an idea that is gaining popularity.[30] Some theorists may think it is right to say that *some* credit is due to an agent whose beliefs are reliably formed, whereas no credit is due to a lucky guesser. But on the other hand, it may seem right to others to *deny* any credit to an agent whose beliefs are reliably formed when she has no inkling of her reliable process, or to an agent who is not motivated to seek truth. The problem is that "credit" seems to be used ambiguously, sometimes entailing proper motivation to seek the truth and sometimes not. Happily, I think there is an alternative, simpler diagnosis of value in a reliabilist anti-luck epistemology. I shall describe a series of cases where an agent has a reliably formed true belief, and where it would indeed be quite controversial to claim that the agent deserves credit for having the belief. Still, I want to say that the agent knows, that her knowledge is better than mere true belief, and that the value derives from her *enviably strong epistemic position*.[31] Thus, while what I say here goes little further than Riggs' defense of the value of reliably formed belief, I shall explicitly discard the explanation of value in terms of credit due to an agent, which partially undermines the impetus for the turn to virtue theory of any stripe, and thus frees us to reconsider the overall merits of generic process reliabilism (as opposed to agent reliabilism).

As I am using the terms, "strong epistemic position" and "reliability" refer to the same thing, roughly, forming beliefs from a process that produces a high ratio of true beliefs in the actual world and in nearby worlds – the more distant the worlds through which one's belief-forming process produces true beliefs, the more reliable the process, hence the stronger the epistemic position. This appears to be both good and bad for my purposes. It is good in that the shift in (mere) terminological focus does not entail abandoning the core reliabilist notion of truth-conduciveness in order to solve the value problem concomitant with it, so I cannot be charged with simply changing the topic. It may be bad because, if the terms refer to the same thing, then if the value problem afflicts reliabilism, it afflicts an account of knowledge cashed out in terms of true belief formed from an enviably strong epistemic position, so what could be gained by appeal to the latter?

Toward an answer to this question, I want to suggest that there is a subtle distinction in connotation between the concept of reliability, or truth-conduciveness, and that of strength of epistemic position, which focuses less on the end than on an agent's capacity for achieving it. The change of focus makes relevant an important distinction in consequentialist ethics that may be applicable, by analogy, to the value problem. Ethicists distinguish between act and rule consequentialism: Both theories construe the goodness of an act in terms of ends, but rule consequentialism recognizes the goodness of a rule even when, in a particular case, acting on it does not achieve the desired ends. With regard to reliabilism, the thought is that we can locate a kind of value in a strong epistemic position even when it issues in a *false* belief. If so, then reliabilists can offer an account of epistemic value that does not hinge solely on whether a belief is true – on whether the end is achieved. Perhaps thinking solely in terms of reliability – mere truth-conduciveness – has blinded us to the possibility that the value of knowledge is not merely in the end of truth but also in the goodness of the belief-forming process.

This is crucial because Zagzebski often invokes the machine-product model of knowledge to raise the value problem:

> The liquid in this cup is not improved by the fact that it comes from a reliable espresso maker. If the espresso tastes good, it makes no difference if it comes from a reliable espresso maker. ... [B]eing the product of a reliable faculty or agent does not add value to the product.
>
> (Zagzebski 2003a, "The search," 13)

Zagzebski is surely correct that if we begin with good espresso and ask whether it would have been better if made by a good espresso maker, it is hard to see how that could be. And though it is clear that having a good espresso maker is itself valuable (which, somehow, seems irrelevant to Zagzebski, whereas I think it is crucial), Zagzebski's complaint is that it does not make a good cup of espresso taste any better than a good cup from a bad machine, just as a true belief formed from a reliable process is no truer than any other true belief. But if the analogy to ethics holds up, it may be that we are asking the wrong question. If a consequentialist rule is valuable even when it does not achieve the relevant ends, then a strong epistemic position may have value even when a belief issuing from it turns out false. Notice that I am not claiming that the false belief is itself valuable, but only that there is surely a source of epistemic value other than truth, even if we think of truth as the only desired end (by analogy, that there is a source of value other than the outcome of a particular act, even if we think of good consequences as the only desired end). On the point that value is not conferred on the resulting belief (whether true or false), there may be a disanalogy with rule consequentialism, one that I will describe later. One important question to consider is whether the disanalogy undermines the case I am making for value. But first things first. Let us look at cases involving a strong epistemic position.

Truetemp world

Keith Lehrer once described the following thought experiment designed to undercut externalist (thus reliabilist) theories of knowledge (TRUETEMP).

> Suppose a person, Mr. Truetemp, undergoes brain surgery by an experimental surgeon who invents a small device that is both a very accurate thermometer and a computational device capable of generating thoughts. The device, call it a tempucomp, is implanted in Truetemp's

> head so that the very tip of the device, no larger than the head of a pin, sits unnoticed on his scalp and acts as a sensor to transmit information about the temperature to the computational system in his brain. This device, in turn, sends a message to his brain causing him to think of the temperature recorded by the external sensor. Assume that the tempucomp is very reliable, and so his thoughts are correct temperature thoughts. All told, this is a reliable belief-forming process and a properly functioning cognitive faculty.
>
> Now imagine, finally, that Mr. Truetemp has no idea that the tempucomp has been inserted in his brain and is only slightly puzzled about why he thinks so obsessively about the temperature; but he never checks a thermometer to determine whether these thoughts about the temperature are correct. He accepts them unreflectively, another effect of the tempucomp. Thus, he thinks and accepts [for example] that the temperature is 104 degrees. It is.[32]

Lehrer argues that Truetemp does not know because he lacks any reason to believe that his beliefs are trustworthy. At best, according to Lehrer, we can say that Truetemp possesses correct *information*, but that is a far cry from knowing the temperature.

In another paper (Becker 2004), I argued that it is at least plausible to say that Truetemp *possesses* knowledge even if one is tempted to say that he does not know. (I went further still and argued that, once one accepts this point, only willful stubbornness allows the anti-externalist to deny knowledge to Truetemp.) But the argument that Truetemp knows the temperature is unnecessary for present purposes. What is necessary is to show that there is a source of epistemic value other than the attainment of a true belief. If we can show this, it rebuts the objection that Truetemp does not know *because* reliabilism cannot account for epistemic value independent of true belief. To that end, I will extend the thought experiment in such a way as to demonstrate how having a reliable belief-forming process is indeed far more valuable than possessing a process whose outputs are in no discernible way connected to the facts that make them true – processes such as guessing, going on hunches, wishful thinking, and the like. Zagzebski's espresso machine analogy obviously invites an inquiry into this kind of value, but in focusing solely on a comparison of a good machine's output to any other good cup of espresso, we are forbidden, arbitrarily in my view, from looking in this promising direction. (If you set up the problem in Zagzebski's way, solely by comparing differently achieved positive ends, you get a difficult problem for reliabilism. Answer? Don't do that. It follows that my account of value is not a *direct* response to Zagzebski's challenge, but (1) I said at the beginning of this section that it is controversial anyway to claim that a particular case of knowledge is itself more valuable than a particular true belief; and (2) the account does explain how being in a position to know is better than merely truly believing.)

Notice that Truetemp's temperature belief-forming process is truth-conducive not only in the actual world, but also in most nearby worlds where his beliefs are caused by the tempucomp. His belief-forming process is thus more stable than the chance mechanism that causes Plantinga's brain lesion patient to have true brain lesion beliefs. Now to extend the thought experiment.

First extension

Suppose that Truetemp's doctor has a very good reason for implanting the thermometer in Truetemp's head. The brain surgery fixes a neurological disorder, but makes Truetemp susceptible to brain damage should the ambient temperature fall outside of some normal specified range, say between zero and 100 degrees Fahrenheit. The doctor says to Truetemp: "If for any reason you believe that the temperature drops below zero or rises above 100, wherever you are, be sure to call me immediately. It may be very important to your health." Still, she does not tell Truetemp anything about the thermometer. Perhaps it was illegal for her to have implanted it.

Now I ask you: Is it better, in some intuitive sense, for Truetemp to have a reliable belief-forming mechanism, even one over which he has no control and of which he knows nothing, or for Truetemp occasionally to guess the temperature correctly? The answer is clearly that the former is far more valuable. Without the accurate thermometer, Truetemp might die.[33] Suppose I had asked: Is Truetemp's particular reliably formed true belief token that it is 104 degrees more valuable than had he merely guessed correctly? The answer to that question is less clear. It seems, then, that focusing on particular instances of reliably and unreliably formed *true* beliefs obscures the added value of having a reliable belief-forming process.

Second extension

Now imagine that, unknown to all but a small group of government officials and brain scientists, everyone in Truetemp world is susceptible to the same debilitating brain disease that afflicts Truetemp. Not wanting to create panic, but recognizing the health requirements of its citizens, every hospital is assigned a brain specialist who performs the relevant surgery and then surreptitiously implants a tempucomp in every newborn. Parents are told not to subject their children to extreme temperatures before the age of three so as to avoid common but serious ailments. This effectively keeps the children's brains safe until they are old enough to report their temperature beliefs. Moreover, each parent is told to contact a doctor if he or she believes the temperature falls outside of the specified range.

The point of the second extension of the thought experiment is to make the artificial tempucomp akin to our natural perceptual cognitive endowments – we are imagining that almost everyone has temperature-sensing abilities just as most have reasonably good vision. Seeing the analogy allows us to bring the added value of reliable belief-forming processes out of the realm of science fiction and into our everyday lives. It is uncontroversial to say that vision-based beliefs often constitute knowledge even absent any special motivation to achieve the truth. When one sees a cat on the mat, one normally believes that there is a cat on the mat straightaway.[34] Of course, an agent who is especially motivated to believe the truth might very well question the veracity of her belief, or the reliability of her vision, but it goes too far to say that this is necessary for knowledge, though it may indicate a virtue of the agent. (On the other hand, it may sometimes indicate paranoia, absent any reason to think one's vision is unreliable, which is not obviously (obviously not?) epistemically praiseworthy.)

It is uncontroversial to say that in the Truetemp cases possessing and using a reliable belief-forming process is more valuable than lucky guessing. It is also clear, however, that the additional value cannot be attributed solely to the outcome, a particular true belief, for truly believing that the temperature exceeds 100 degrees might save one's life whether reliably formed or not. The question, then, is: Whence the added value of *reliably formed* true belief?

My answer is simple and straightforward: Truetemp is in an *enviably strong epistemic position*; that it is enviable indicates that it is valuable, even though Truetemp deserves no credit for possessing the reliable process – for being in that position. (One might think here of the person who gets the promotion because of nepotism, and is therefore envied but deserves no credit. Note too that just because Truetemp's epistemic position is envi*able*, it does not entail that anyone actually envies it.) There are different ways of assessing the strength of one's epistemic position, including appeal to modal principles such as sensitivity and safety, which, we have seen, apply to particular beliefs as opposed to belief-forming processes. Since my main aim here is to defend generic *process* reliabilism, those kinds of principles are irrelevant. For present purposes, a strong epistemic position is one such that most beliefs formed from it are true in the actual world and in nearby worlds. In Truetemp's case, it is easy to see that his true beliefs, unlike guesses, would be true throughout many (perhaps most) nearby worlds. This is really just stipulated in the description of the case. If there is no tempucomp and he merely guesses, even if he is right in the actual world, he could easily go wrong in nearby worlds.

One nice thing about locating the value of reliably formed true belief in an enviably strong epistemic position is that it is not debatable whether an agent having a process that is modally (or counterfactually) reliable *entails* that the agent is in a strong epistemic position. One could easily argue that Truetemp deserves no credit for his true beliefs, since he knows nothing about the tempucomp. But to claim that he is not in a strong epistemic position just is to deny that his belief-forming process is modally reliable, which is not an option.

To pinpoint the value of reliability, we must follow Riggs (2002, 87) in rejecting the twin assumptions in the case against reliabilism. One is that the *only* value of reliably formed true belief is in the actual outcome, and the other is that value accrues to the belief itself. The former must be wrong because it implies that the only relevant good is attaining a true belief, and if that is all that matters, it would seem irrelevant how one achieves it. The latter must be wrong for similar reasons. If the value must be "in" the belief, and the only value of a belief itself is truth, then one kind of true belief is no better than another. However, we do not want to go so far as to say that value derives from credit due to an agent. (Again, even if some believe this is right, it is extremely controversial to claim that the agent, for instance Truetemp, deserves credit. It is not controversial to say that he is in a strong epistemic position.)

The value of (being in) a strong epistemic position is neither merely in its outcome nor simply a feature of the belief itself. It is true that strength is measured by truth-conduciveness, hence the applicability of the value problem, but further value is indicated by the obviously enviable position of the agent. The point seems rather obvious to me, but maybe that is just because I made it. Before further articulating the value of an enviably strong epistemic position, let me offer an analogy involving a strong, that is, valuable position outside the scope of any philosophical issues.

Smith and Jones are playing BLACKJACK. Smith is dealt 6–5, and Jones is dealt 4–8; the dealer is showing a 5. Smith and Jones care about nothing except the outcome, namely, winning money. At this point, Smith is in a very strong position, while Jones is in a weaker position. Smith doubles down. That is, she doubles her bet because the goal is to get as close to 21 as possible, without going over, while having a better hand than the dealer, and the dealer is showing a "bust card," one with a greater likelihood of making the dealer go over 21 and thus having to pay all the players who don't bust. In exercising the double-down option, Smith is allowed to receive only one more card. Suppose that Smith draws a 4, for a total of 15. Jones, who does not really grasp the game, also doubles down, luckily draws an 8, and thus has a hand totaling 20. The dealer flips her down card, a 4, and draws a king for a total of 19. Jones wins twice her original bet; Smith loses twice her original bet.

How should we describe this example? First, the only thing any player cares about is winning money. Second, Smith and Jones received their first two cards at random, and so deserve absolutely no credit for their respective positions. Third, Smith forms her ultimate hand from an enviably strong blackjack position, while Jones forms hers from a weaker position. Finally, only Jones wins money, the ultimate end.

What this analogy suggests is that there is value in an enviably strong position even when it does not deliver the desired end. This, in turn, suggests that we lose sight of the value when we focus solely on the end in a particular case – winning a lot of money or having a true belief – and ask: What value is added to these achieved ends by being in an enviably strong blackjack position, or having a reliable belief-forming process? To see the value derived from the difference in kind between positions, we might instead ask: Even if the end had not been secured, was there something of value in the position? (Even if the end is not secured, is there something of value in the position?) The answer is yes, and it carries over to the Truetemp case. Even if, in a particular instance, Truetemp is wrong about the temperature – modal reliability does not guarantee truth – we recognize the value of his epistemic position vis-à-vis that of a lucky guesser. We would *rather be* in his position.

I have been arguing, partly by analogy (and following Riggs's lead), that the value of knowledge is not in the belief itself, but in possessing a reliable process. I have been concerned, however, that explaining this value in terms of credit due to an agent, whether to mollify motivational virtue theorists who deem praiseworthiness an essential ingredient in knowledge of for some other reason, is at best contentious, for instance in cases like Truetemp or even in many ordinary instances of perceptual knowledge, and at worst obviously false. I therefore substituted the idea that reliable belief-forming processes, in producing beliefs that are not merely accidentally true, indicate enviably strong epistemic positions rather than creditworthiness. Finally, I suggested earlier that an analogy with the distinction between act and rule consequentialism can help us to see this point.

I also indicated, however, that there is a disanalogy with the ethics case, and I want to close on this issue. The worry, one might argue, is that rule consequentialists see value in the general rule, even when acting on it fails to achieve the desired results, precisely because the agent is motivated to act on the rule. There are two related potential problems here. First, motivation has entered the picture here, and with it a basis for praiseworthiness. Without these features, is there anything of worth left in the analogy? Second, the motivational element in rule consequentialism implies that a *particular act* is itself more valuable when based on and so motivated by a good rule, whether it achieves the desired results or not, than one not so motivated. That is, the act can derive value from the rule on which it is based, such that an act with good (or bad) consequences that is right rule-based is deemed better than one not so based. I have argued, in the spirit of a non-motivational process reliabilist account of epistemic value, and in agreement with Zagzebski's complaint with the machine-product model, that the process does not confer value (independently of truth) on particular beliefs, just as being in a strong blackjack position does not make a winning (or losing) hand any more valuable than being in a weak position. Again, what remains of the analogy?

Only an internalist or otherwise responsibilist theorist of knowledge would think of the motivational aspect of the *analogy* essential. To say this is not to deny that it is an important feature of rule consequentialist ethical theory *independently* of the analogy, but the salient feature for analogical purposes is that *possessing* the rule is valuable even in cases where acting on it does not result in the desired consequences. That is, the point of the analogy is to identify a source of value within a broadly consequentialist theory that is independent of motivation. It thus goes too far to reply against the externalist that it is not the *right kind* of value because, for instance, in the ethical rule consequentialist framework, the *real* value is the motivation. The challenge was to show that there is a source of value in reliability, and we found it in its concomitant enviably strong epistemic position.[35] Similarly, even without its motivational aspect, there is value in possessing a good rule of action at least to the extent that it increases the likelihood of good consequences, and thereby decreases the chances that good consequences are mere accidents. Imagine that good rules were programmed into us such that we act on them automatically. If we care only about good consequences, for instance overall happiness (or truth), this would be a better world than one where acts (or beliefs) yield overall happiness (or truth) only randomly. Admittedly, this is not much of a basis for an ethical theory, but it does reveal a source of value. The anti-reliabilist is free to reject reliabilism on some or other grounds, but not because it cannot provide for epistemic value independent of attaining a true belief.

C. Conclusion

I saved my treatment of the value problem and of the worry about the counterintuitiveness of the denial of closure for this last chapter because, as I see it, solving these problems is not compulsory. Some philosophers believe that it is simply impossible that the closure principle could be false, and others believe that cases of knowledge are inherently more valuable than mere true beliefs. I laid out a theory of knowledge in Chapter 5 that handles just about every difficult case one could throw at it. If it succeeds and entails both non-closure and some version of the value problem, then a reasonable reply would be: So much the worse for closure and the putative value of knowledge over mere true belief.

On the other hand, it is clear from the recent literature that such a reply may be taken as, at best, evasive and, at worst, haughty. So I have tried to show that reliabilism explains in a plausible way *how* closure fails, not merely entailing *that* it does, while at the same time giving due respect to the powerful appeal of skeptical worries. I have also worked to show that there is a source of epistemic value in reliabilism that does not rely on the actual attainment of truth. All I ask, in concluding this chapter and book, is that you avoid the temptation to use the standard philosophical tools in rejecting the results from this chapter. Nothing here is meant to be an argument in favor of reliabilism – that work was completed in Chapter 5. Hence the complaint that the arguments here are question-begging – that they work only if we assume the truth of reliabilism – is unfair. I *have* assumed that reliabilism is true, precisely on the strength of all the considerations leading up to this chapter. I only wanted to explain, first, how we can live comfortably with the supposed negative consequence that the version of reliabilism favored herein denies closure, and second, that we need not accept the supposed negative consequence that there is no more value when one is in a knowing position than when one merely truly believes.

Notes

1 Introduction: Externalism and modalism

1 It is not always obvious which of the three strategies a theorist is using. Armstrong argues that the Gettier examples do not work because they hinge on an agent having false propositions among her justificatory grounds, and thus that her beliefs are not in fact justified. David Armstrong, *Belief, Truth, and Knowledge* (Cambridge: Cambridge University Press, 1973), 152. One who follows this line of thought may be seen as amending the notion of justification, claiming that it may include no false grounds. Lehrer and Paxson maintain that there can be no defeaters to one's justification, which falls squarely within the strategy of adding a fourth condition. On their view, knowledge is undefeated, justified true belief. Keith Lehrer and Thomas Paxson, "Knowledge: undefeated justified true belief," *Journal of Philosophy* 66:8 (1969): 225–37.

2 It is not even clear that there is a univocal, traditional concept of justification. Alston's recent book enjoins us to dispense with the notion of justification entirely in our theorizing because "There isn't any unique, epistemically crucial property of beliefs picked out by 'justified'." William Alston, *Beyond Justification* (Ithaca: Cornell University Press, 2005), 22; hereafter cited in text as Alston (2005). Instead, theorists who seem to disagree about the nature of justification are simply emphasizing different valuable aspects of belief. I'm inclined to agree with this assessment, and I shall say more about my reasons for not pursuing an account of the belief-truth link in terms of justification, below.

3 Jason Stanley, *Knowledge and Practical Interests* (Oxford: Oxford University Press, 2005), 114.

4 It would not be wrong here, were one to anticipate the theory I will defend, to complain that I, too, am ignoring a deeply held intuition, namely, that the closure principle is true. I admit it. But if the cost of accepting closure, a principle which, as is often pointed out, is *not* logically valid, is either to acquiesce to skepticism, a doctrine that almost nobody believes, or to lay claim to knowledge that radical skeptical hypotheses are false – which, were that correct, would make it difficult to understand how skepticism was ever a concern – I think it's reasonable to take the denial of closure seriously.

5 Or culturally relative. A concern recently expressed by Bishop and Trout is that contemporary epistemology is not much more than a description of analytic epistemologists' intuitive judgments, and those judgments may be unreliable. They report a study involving a Gettier-style example, where subjects were asked to say whether the agent "REALLY KNOWS" or "ONLY BELIEVES." Michael Bishop and J.D. Trout, "The pathologies of standard analytic epistemology," *Noûs* 39:4 (2005): 696–714, p. 705. Many East-Asian subjects said, in contrast to

the analytic epistemologists' intuitions, that the subject really knows. But surely this is a false dichotomy. What about "ONLY HAS A TRUE BELIEF"? Had that option been available, it might have given Asian subjects what they were looking for.

6 Keith DeRose's well-known work features a sensitivity element, but he does not claim that it is a necessary condition for knowledge. See Keith DeRose, "Solving the skeptical problem," in Keith DeRose and Ted Warfield, eds. *Skepticism: A Contemporary Reader* (Oxford: Oxford University Press, 1999), 183–219; hereafter cited as DeRose (1999). (Originally in *Philosophical Review* 104:1 (1995): 1–52.) See also Chapter 6 of this book.

7 And it may or not be, but I will not even bother to cite any particular works of the subsequent literature in favor of or against the view that externalism about meaning is reconcilable with privileged knowledge of one's propositional attitudes. The literature is just too vast. A place to get your feet wet is Susana Nuccetelli's edited volume, *New Essays on Semantic Externalism and Self-Knowledge* (Cambridge, MA: MIT Press, 2003), which has a very good bibliography. Note too that an individualistic individuation of mental states does not make the problem of self-knowledge disappear. For example, if such states are individuated solely by reference to the thinker's brain states (as opposed to features of the environment external to the agent), it is not as straightforward as individualists often seem to suggest that one could know them in a privileged way. For more on this, see my "Individualism and self-knowledge: *Tu Quoque*," *American Philosophical Quarterly* 39:3 (2002): 289–95.

8 Hilary Putnam, "The meaning of 'meaning'," in *Mind, Language and Reality* (Cambridge: Cambridge University Press, 1975), 215–71.

9 Tyler Burge, "Individualism and the mental," in Peter French, Theodore Uehling, Jr., and Howard Wettstein, eds. *Midwest Studies in Philosophy* 4, *Metaphysics* (Minneapolis, MN: University of Minnesota Press, 1979), 73–121.

10 There are good reasons for doubting whether Wittgenstein sought this conclusion, rather than the less dramatic conclusion that there is no private language that nobody but the "private linguist" could understand. (For Kripke's interpretation, see his *Wittgenstein on Rules and Private Language* (Cambridge, MA: Harvard University Press, 1982).) At any rate, here I am just sketching externalism about meaning. The claim that Quinean indeterminacy leads to a communal view of the determinants of reference (and thus meaning, on the assumption that meaning determines reference) is less straightforward, but in "Ontological relativity," Quine claims that questions about the referents of an individual's terms are settled by "acquiescing in our mother tongue and taking its words at face value," which suggests, prima facie at least, a communal view. W.V. Quine, "Ontological relativity," in *Ontological Relativity and Other Essays* (New York: Columbia University Press, 1969a), 26–68. p. 49.

11 This is a pretty standard internalist view and can be found in Chisholm: "the things we know are justified in the following sense: *we* can know what it is, on any occasion, that constitutes our grounds, or reason, or evidence for thinking that we know." Roderick Chisholm, *Theory of Knowledge*, 2nd edition (Englewood Cliffs, NJ: Prentice-Hall, 1977), 17. It is defended by many; a version that considers some of the newer developments in epistemology can be found in Matthias Steup, "A defense of internalism," in Louis Pojman, ed. *The Theory of Knowledge*, 3rd edition (Belmont, CA: Wadsworth, 2003), 310–21; hereafter cited in text as Steup (2003). When discussing internalism explicitly, I shall often use the term "justification" for the requisite link between belief and truth. When discussing externalism, I often use "warrant," which may be understood as a more general term.

12 Alvin Goldman, "A causal theory of knowing," *Journal of Philosophy* 64:12 (1967): 355–72; hereafter cited as Goldman (1967).

13 Direct realist views of perceptual knowledge are compatible with externalism in this sense, since they reject the view that perceptual belief is mediated and hence justified by some immediately knowable ground. For a defense of this view against the standard argument from illusion, see William Alston, "Perceptual knowledge," in John Greco and Ernest Sosa, eds. *The Blackwell Guide to Epistemology* (Oxford: Blackwell, 1999), 223–42.

14 For instance, Guy Axtell argues, for well-motivated reasons, that justification involves both internalist and externalist elements and thus abandons "pure" internalism and externalism in favor of a "mixed" view. Guy Axtell, "Felix Culpa: luck in ethics and epistemology," *Metaphilosophy* 34:3 (2003): 331–52, pp.340–42; hereafter cited as Axtell (2003). But any view according to which what warrants belief may involve factors not recognizable upon reflection is externalist. Axtell is certainly not confused about this: "Such accounts are externalist in the standard sense that not all the constraints they place on justification need to be internally accessible" (341).

15 Greco endorses the claim that internalism is best construed as the thesis that "the epistemic status of a belief is *entirely* determined by factors that are relevantly 'internal' to the believer's perspective." John Greco, "Justification is not internal," in Matthias Steup and Ernest Sosa, eds. *Contemporary Debates in Epistemology* (Oxford: Blackwell, 2005), 257–70, p. 257 (my emphasis).

16 I doubt that even this much "internalism" is required for knowledge. Again, were one to tell the story about an agent's knowledge of Lincoln's birth year carefully, it is plausible to say that she knows with absolutely no internally accessible grounds. She may not remember where she read it or even that she read it, and she may not even have a clue as to why she believes it. Still, her in fact having read it in a reliable source and remembered it is reason enough, at least for me, for thinking she knows it. If you've ever watched *Jeopardy!* and played along, you've probably been surprised at some of the things you know without having any idea where you picked them up.

17 Content externalism appears to be in tension with justification internalism, since one is internally justified only if one has access to one's grounds, but the grounds themselves, in the form of supporting believed propositions, are individuated by reference to features of the environment about which the agent may be ignorant. James Chase attempts to square content (meaning) externalism and epistemological internalism. James Chase, "Is externalism about content inconsistent with internalism about justification?" *Australasian Journal of Philosophy* 79:2 (2001): 227–46.

18 Now that I have described externalism more fully, I would like to echo another point made by Alston in his recent book. (See also note 2, above.) When one thinks of the link between truth and belief that constitutes knowledge in truth-conducive terms, as externalists typically do, "there is something of a linguistic strain in speaking of justification in these terms" (Alston 2005, 16). Reversing the point, if one then continues to use the term "justification" for this link, one inevitably encounters the internalist criticism that he or she is not adequately characterizing the concept. This is part of what I mean by "preconceptions loaded into the notion of justification."

19 Well, Aristotle is sometimes thought of as an epistemological externalist, so I guess it's not all that recent! Even so, epistemology since the early modern era has been dominated by internalism. On the point about over-intellectualization, I think a similar thing could be said about the history of the theory of meaning. (It *has* been said by Burge, most recently in his 2005 Carus lectures.) Enter externalism: One may be competent with the use of an expression, hence possess a concept, without thereby knowing all of, or fully understanding, its individuation conditions.

20 This is not to denigrate Leibniz, but discussion of possible worlds in wider philosophical contexts was more or less dormant in the interval from Leibniz to Kripke.
21 David Lewis, *Counterfactuals* (Oxford: Blackwell, 1973).
22 It is interesting that mathematical entities have a similarly problematic ontological status – what and where are numbers? – but nobody seems to think that this renders math talk *unclear*. Of course, math talk plays a more central role in our world view than possible worlds discourse, and so nobody would dream of dispensing with it. The present suggestion is that possible worlds talk can be similarly useful, as least for philosophical purposes.
23 Here is Richard Fumerton making a similar point in criticizing Nozick's use of possible worlds to explicate subjunctives: "God (or Kripke) only knows how far away from the actual world a possible world must be in order to be irrelevant to the truth of a subjunctive conditional." Richard Fumerton, "Nozick's epistemology," in Steven Luper-Foy, ed. *The Possibility of Knowledge: Nozick and His Critics* (Totowa, NJ: Rowman & Littlefield, 1987), 163–81, p. 167.
24 Thanks to both Brian Cooney and John Greco for getting me to place the focus here on the usefulness of the appeal to possible worlds that underlies counterfactual and subjunctive conditionals.
25 Alvin Goldman, "What is justified belief?" in George Pappas, ed. *Justification and Knowledge* (Dordrecht: D. Reidel, 1979), 1–23, p. 10.
26 Alvin Goldman, whose work is the focus of the following chapter, uses the terminological distinction between global and local, which he borrows from McGinn. See Alvin Goldman, *Epistemology and Cognition* (Cambridge, MA: Harvard University Press, 1986), 44ff. and Colin McGinn, "The concept of knowledge," in Peter French, Theodore Uehling, Jr. and Howard Wettstein, eds. *Midwest Studies in Philosophy* 9, *Causation and Causal Theories* (Minneapolis, MN: University of Minnesota Press, 1984), 529–54. Jessica Brown also follows this terminology. *Anti-Individualism and Knowledge* (Cambridge, MA: MIT Press, 2004), 120ff. Jonathan Vogel, to take another example, labels reliabilism with regard to particular beliefs, such as sensitivity – what we are calling local reliabilism – "counterfactual reliabilism." See his "Reliabilism leveled," *The Journal of Philosophy* 97:11 (2000): 602–23.
27 Duncan Pritchard, "Virtue epistemology and epistemic luck," *Metaphilosophy* 34:1/2 (2003): 106–30, p. 111; hereafter cited in text as Pritchard (2003).
28 Riggs offers a different taxonomy of luck, but one that, very roughly speaking, maps onto Pritchard's. See Wayne Riggs, "What are the chances of being justified?" *The Monist* 81:3 (1999): 452–72. Riggs's general notion is chance, which he divides into luck and accidents, where luck is fairly close to Pritchard's veritic luck, and accident is pretty close to reflective luck. Engel uses the term "veritic luck" for the kind of luck where one's belief just happens to be true even though it is justified. Mylan Engel, "Is epistemic luck compatible with knowledge?" *The Southern Journal of Philosophy* 30:2 (1992): 59–75. He distinguishes this from "evidential luck," roughly, Pritchard's reflective luck, and argues that *only* veritic luck is incompatible with knowledge. Hall rejects Engel's claim that *even* veritic luck is incompatible with knowledge. Barbara Hall, "On epistemic luck," *The Southern Journal of Philosophy* 32 (1994): 79–94. Vahid is also a compatibilist who introduces an essential modal element: "We can, thus, further distinguish between knowledge-consistent luck and knowledge-precluding luck, a distinction which is a direct function of how easily the cognizer's belief could have changed its status." Hamid Vahid, "Knowledge and varieties of epistemic luck," *Dialectica*, 55:4 (2001): 351–62, p. 357. For one of the earliest discussions of epistemic luck, see Richard Foley, "Epistemic luck and the purely epistemic," *American Philosophical Quarterly* 21:2 (1984): 113–24. The soon-to-be *locus*

classicus on epistemic luck is Pritchard's *Epistemic Luck* (Oxford: Oxford University Press, 2005); hereafter cited as Pritchard (2005).

29 I discuss virtue epistemology, or more precisely, I respond to the value problem for reliabilism raised by virtue epistemologists, in Chapter 6. Responsibilist versions claim that a belief is justified only if the agent does her epistemic duty in forming belief. See, for instance, Jonathan Kvanvig, *The Intellectual Virtues and the Life of the Mind: On the Place of Virtues in Epistemology* (Lanham, MD: Rowman and Littlefield, 1992); James Montmarquet, *Epistemic Virtue and Doxastic Responsibility* (Lanham, MD: Rowman and Littlefield, 1993); and Linda Zagzebski, *Virtues of the Mind* (Cambridge: Cambridge University Press, 1996); hereafter cited as Zagzebski (1996). Kvanvig's later book expresses some skepticism about any forthcoming explanation of the value of knowledge, but argues that an account of other epistemic achievements, especially understanding, provides a more promising basis for an explanation of their particular kinds of value. See his *The Value of Knowledge and the Pursuit of Understanding* (Cambridge: Cambridge University Press, 2003).

30 I am being extremely gracious here in even suggesting that it is my externalist commitment that leads me to say the chicken sexer knows, when in fact the latter kind of belief led me to externalism. Axtell (2003, 340) claims that externalists ignore the problem of reflective luck and internalists that of veritic luck only because of the logics of their theories. For what it's worth, I intend my basic beliefs about cases of knowledge to determine my theoretical commitments, and not vice versa. That is why my change of mind resulting from the Abraham Lincoln case was such an eye-opener.

31 Edmund Gettier, "Is justified true belief knowledge?" *Analysis* 23:6 (1963): 121–23.

32 This is essentially the Haveit case that I will (re)introduce in Chapter 3.

33 I am calling it process luck because I will ultimately argue for a process reliability condition on knowledge. But at this early stage it would be unfair to force the discussion in that direction. Thus if you believe, for instance, that knowledge requires belief formed from a stable cognitive disposition, you might call the phenomenon I describe here something like *cognitive dispositional luck*.

34 Process luck, then, exists on a continuum, as does process reliability. Eliminating process luck requires a condition where the threshold ratio of false belief outputs is relevantly low, hence where the threshold ratio of true belief outputs is adequately high. My claim is that the theory must eliminate process luck, but that there is no precise ratio of false-to-true beliefs that constitutes process luck. Still, there are cases where process luck is clearly at work, and this is because it is plausible to say that the proportion of false beliefs produced by the process is too high.

35 For an excellent discussion of the importance of skeptical arguments and how they are not at all easily dismissed, see John Greco, *Putting Skeptics in Their Place* (Cambridge: Cambridge University Press, 2000). The preface by itself is worth the price of the book; it is illuminating and answers very closely to experiences I have had in discussing skepticism with philosophers who reject skeptical worries out of hand.

36 Christopher Hill, "Process reliabilism and Cartesian scepticism," *Philosophy and Phenomenological Research* 56:3 (1996): 567–81.

37 Ernest Sosa, "Philosophical scepticism and epistemic circularity," *Aristotelian Society Supplementary Volume* 68 (1994): 263–90.

38 Barry Stroud, "Scepticism, 'externalism', and the goal of epistemology," *Aristotelian Society Supplementary Volume* 68 (1994): 291–307.

39 Stroud would like to dig in his heels here, if possible, and say that there is something wrong about entertaining the skeptic's global hypothesis. But this, in my view, errs in the opposite direction of simply not taking skepticism seriously

enough. Without some principled reason for denying the legitimacy of skeptical scenarios, ignoring them is tantamount to plugging one's ears. Some may find such a principled reason in Wittgenstein's *On Certainty*, where Wittgenstein claims that some propositions, for instance, Moore's *I Have Two Hands*, are hinges upon which our scientific (and, more generally, empirical) investigations swing, and thus that we ought not doubt them – to doubt them is to give up the game of empirical inquiry altogether. (See Wittgenstein, *On Certainty* (Oxford: Blackwell, 1969), §341ff.) But that does not mean that such beliefs have a positive epistemic status and thus constitute knowledge of either the hinge propositions or the denials of the skeptical hypotheses that conflict with them, and so the anti-skeptical consequences of the hinge proposition approach are unclear. For a nice, brief discussion, see Pritchard, "Recent work on radical skepticism," *American Philosophical Quarterly* 39:3 (2002b): 215–57, §7.

40 This remark demands closer attention than I can give it here. The internalism/externalism debate concerns the location and cognitive accessibility of factors that warrant belief. As such, it does not obviously apply to many contextualist theories, which, as we will see, are primarily theses about the context-sensitivity of knowledge attributions rather than warrant. Annis (1978) is a clear exception, because he claims that justification is context-sensitive. David Annis, "A contextualist theory of epistemic justification," *American Philosophical Quarterly* 15 (1978): 213–19. Cohen also suggests that justification is context-sensitive, and his view of justification is broadly internalist. Stewart Cohen, "Contextualism, skepticism, and the structure of reasons," *Philosophical Perspectives* 13 (1999): 57–89; hereafter cited as Cohen (1999).

41 For the first variety, see David Lewis, "Elusive knowledge," *Australasian Journal of Philosophy* 74:4 (1996): 549–67, and DeRose (1999). Instances of the second variety are found in Cohen (1999) and Cohen, "Contextualism and skepticism," *Philosophical Issues* 10 (2000): 94–107.

42 Linda Zagzebski, "From reliabilism to virtue epistemology," in Guy Axtell, ed. *Knowledge, Belief, and Character* (Lanham, MD: Rowman and Littlefield, 2000), 113–22.

2 Process reliabilism

1 As I indicated in the introduction, I would prefer not to use the term "local reliabilism" at all, since it makes no essential reference to the truth-conduciveness of a belief-forming process's outputs, taken as a whole. But I acquiesce in the interest of consistency with others' usage.

2 Alvin Goldman, "A causal theory of knowing," *Journal of Philosophy* 64:12 (1967): 355–72; hereafter cited as Goldman (1967).

3 See W.V. Quine, "Epistemology naturalized," in *Ontological Relativity and Other Essays* (New York: Columbia University Press, 1969b), 69–90.

4 I thank Alvin Goldman for pointing out to me this implication of his causal theory.

5 An early critic of the adequacy of Goldman's account of inferentially based empirical knowledge is Harman. He offers examples to show that one may know even when one *incorrectly* reconstructs the relevant causal sequence, from which he concludes that Goldman's causal account is too restrictive. Gilbert Harman, *Thought* (Princeton: Princeton University Press, 1973), 126–35. Goldman later offers an analogous example. Alvin Goldman, *Epistemology and Cognition* (Cambridge, MA: Harvard University Press, 1986), 42–4; hereafter cited as Goldman (1986). I shall focus, however, on cases where the causal requirement is too permissive.

6 Alvin Goldman, "Discrimination and perceptual knowledge," *Journal of Philosophy* 73:20 (1976): 771–91, pp. 772–73; hereafter cited as Goldman (1976).

7 A fourth class is suggested by recent independent work of Hawthorne and Stanley, according to which whether an agent knows depends on the counter-possibilities that are salient to *her*, rather than to knowledge attributors. See John Hawthorne, *Knowledge and Lotteries* (Oxford: Clarendon Press, 2004), and Jason Stanley, *Knowledge and Practical Interests* (Oxford: Oxford University Press, 2005). This view is often called subject sensitive invariantism, but Wright dubs it subject contextualism because, like attributor contextualism, attitudes, intentions, and purposes are salient to one's knowledge status. Crispin Wright, "Contextualism and scepticism: even-handedness, factivity and surreptitiously raising standards," *The Philosophical Quarterly* 55:219 (2005): 237–62. In cases of self-attribution, it yields the same verdict as attributor contextualism about the agent's knowledge status.

8 Fred Dretske, "Epistemic operators," *Journal of Philosophy* 67:12 (1970): 1007–23; hereafter cited as Dretske (1970).

9 Gail Stine, "Skepticism, relevant alternatives, and deductive closure," *Philosophical Studies* 29 (1976): 249–61.

10 It follows that it may be impossible to *express* that knowledge, since expressing it is to raise the skeptical possibility, thus making it relevant. I return to this problem in Chapter 6.

11 In fact, one bit of evidence that Goldman denies the contextualist approach to relevant alternatives is that he appears happy to deny that knowledge is closed under known entailment. (See, for instance, Goldman, "Psychology and philosophical analysis," *Proceedings of the Aristotelian Society* 89 (1989): 195–209.) As we have just seen, a central motivation for contextualism is to uphold closure. Goldman's acceptance of the possibility of closure failure at least suggests that he does not have this central contextualist motivation, and thus that, for him, relevance is not determined by attributors' intentions or purposes or conversational context.

12 For the revised view, see Dretske, "Conclusive reasons," *Australasian Journal of Philosophy* 49:1 (1971): 1–22; hereafter cited as Dretske (1971).

13 Alvin Goldman, "What is justified belief?" in George Pappas, ed. *Justification and Knowledge* (Dordrecht: D. Reidel, 1979), 1–23, p. 11; hereafter cited as Goldman (1979).

14 The idea that there is better and worse knowledge is gaining steam. See, for example, Stephen Hetherington, *Good Knowledge, Bad Knowledge* (Oxford: Clarendon Press, 2001). I am not suggesting, though, that Hetherington would accept my claims about subjunctive conditionals. In fact, he thinks that true belief alone constitutes minimal knowledge.

15 Pritchard also raises the worry that sensitivity takes into account too few nearby worlds, and adds that, if we read sensitivity such that it demands true belief throughout more counterfactual worlds than the nearest one, it is no better, and in some respects worse, than the safety principle. Duncan Pritchard, *Epistemic Luck* (Oxford: Oxford University Press, 2005), 158–59. We will look closely at safety in Chapter 4 and will then be able to see more clearly the differences between safety and sensitivity.

16 A version of the actual world reliabilist view can be found in Colin McGinn, "The concept of knowledge," in Peter French, Theodore Uehling, Jr. and Howard Wettstein, eds. *Midwest Studies in Philosophy* 9, *Causation and Causal Theories* (1984), 529–54.

17 Alvin Plantinga, *Warrant and Proper Function* (New York: Oxford University Press, 1993), 199.

18 See Goldman (1979, 16–17). For a similar example, see John Greco, *Putting Skeptics in Their Place* (Cambridge: Cambridge University Press, 2000); hereafter cited as Greco (2000).

19 Whether it might rather be characterized as an example of veritic luck (which my definition of process luck excludes – see the introduction, §C.2), hinges on the issue of interpreting subjunctive conditionals that we just discussed. Suppose my wish-based belief that the Twins beat the Yankees is true because the demon makes it so. On the pure subjunctives view, if it were false, I would not believe it, and so one might say that my belief would not easily have been false, and thus would *not* be a case of veritic luck but *would* be a case of process luck. On the other hand, since in many nearby worlds the demon does not help at all, one could say that if my belief were false, I would believe it anyway. This would motivate calling it a case of veritic luck.

20 See Stewart Cohen, "Justification and truth," *Philosophical Studies* 46:3 (1984): 279–95.

21 See his "Strong and weak justification," in James Tomberlin, ed. *Philosophical Perspectives* 2 (1988): 51–69, §VIII; hereafter cited as Goldman (1988).

22 And, in case it is not clear yet, I mean the trap set by (1) the claim that justification is necessary for knowledge and (2) the need for an account of the intuitive notion of justification (confused and equivocal though it may be). I should point out that Goldman no longer accepts the normal worlds view of reliability. "On the present rendering, it looks as though the folk notion of justification is keyed to dispositions to produce a high ratio of true beliefs in the actual world, not in 'normal worlds'." Alvin Goldman, "Epistemic folkways and scientific epistemology," in *Liaisons: Philosophy Meets the Cognitive and Social Sciences* (Cambridge, MA: MIT Press, 1992), 155–75, p. 164. As we have seen, however, legitimate reliability requires truth-conduciveness in nearby worlds, too, a point I do not think Goldman denies (see below).

23 The intuition that Goldman aims to capture here seems to be that, to be truly justified, there must be no defeaters to one's justification. For other versions of this idea, see Lehrer and Paxson, "Knowledge: undefeated justified true belief," *Journal of Philosophy* 66:8 (1969): 225–37, and Moser, *Knowledge and Evidence* (Cambridge: Cambridge University Press, 1989).

24 Brian Cooney pointed out to me that it is surely a stretch to say that the parents are trustworthy. He's right, but since the general thrust of the discussion here is that an adequate analysis of knowledge need not (and, in my view, should not) include an account of the traditional notion of justification, Cooney's point is easy for me to accept.

25 On this I seem to be in agreement with Dretske. (See his 1971, n. 17.)

26 Sosa points out correctly that this is an extremely weak sense of "blameless," and he tries to rectify Goldman's characterization. Ernest Sosa, *Knowledge in Perspective* (Cambridge: Cambridge University Press, 1991), 133–4.

27 Modalized process reliabilism should be distinguished from George Bealer's more narrowly circumscribed "modal reliabilism," which is his view that there is a modal tie between a priori (evidential) intuition and truth. George Bealer, "A theory of the a priori," *Pacific Philosophical Quarterly* 81 (2000), 1–30. If Bealer is right, then a priori beliefs formed on the basis of intuitions would constitute knowledge under the modalized process reliabilism rubric, but would (obviously) not exhaust the latter category. Incidentally, another point of contact between Bealer's work on intuition and the present work is that the former may be used to justify the basic methodology used here. One's intuitions about cases of ϕ are modally reliable insofar as one's concept of ϕ is (what Bealer calls) semantically stable (*ibid.*, 19). The suggestion is that widespread agreement in intuition about cases of knowledge implies that the concept of knowledge is at least well enough understood and sufficiently stable to be appropriately modally tied to the truth.

28 See, for example, Greco (2000) (Chapters 7 and 8), Greco, "Virtue and luck, epistemic and otherwise," *Metaphilosophy* 34:3 (2003a): 353–66, and Sosa (1991).

29 Linda Zagzebski, *Virtues of the Mind* (Cambridge: Cambridge University Press, 1996), 285ff.

30 Dretske (1971, 16–17) offers a similar case, but one involving a *dog* being fooled into thinking a thief is his owner, while the owner lies out cold in a corner.

31 Zagzebski's specific target here is Plantinga's (1993) proper function account of warrant. But as Pritchard neatly points out, the example undermines many kinds and relatives of process reliabilist theories. Duncan Pritchard, "Virtue epistemology and epistemic luck," *Metaphilosophy* 34:1/2 (2003): 106–30. Mary's vision-based belief is formed by a stable disposition and a reliable character trait, in Sosa's sense, where "character trait" does not necessarily imply a trait for which an agent is responsible.

32 Why not revert to the original causal account? We have already seen that it is too weak, and then there is the problem of knowledge of necessary truths – how can they be causes of belief? Something more general is called for. I should point out that it is not entirely clear how the Mary case is any different for present purposes than the barn facsimile case. Henry's eyesight, we may suppose, is exceptional. The barn facsimile case is even stronger, more difficult, in that Henry is in fact looking at a real barn, while Mary is looking at her husband's brother. Though Goldman (1986) does insist on the need for both local and global reliabilist elements in a theory of knowledge, even the relatively early barn case – one of the original motivations for Goldman's development of process reliabilism – calls for a local reliabilist requirement. Goldman's answer to the barn facsimile case provides just that, but it is not obvious whether Goldman is primarily interested in local or global reliabilism in the "Discrimination" paper. No matter. Here we just need to be straight on our own problems and theoretical requirements, and the Mary case forces us to confront the limitations of process reliabilism.

33 Goldman (1979, 12) discusses this. See also Feldman, "Reliability and justification," *The Monist* 68:2 (1985): 159–74, and Conee and Feldman, "The generality problem for reliabilism," in Louis Pojman, ed. *The Theory of Knowledge*, 3rd edition (Belmont, CA: Wadsworth, 2003), 296–309, for definitive statements of the problem.

34 Individuating processes very narrowly is also an idea once suggested by Goldman, but I am not aware of any one place where he puts this together with the content-neutrality idea to solve the generality problem.

35 See Christopher Hill, "Process reliabilism and Cartesian scepticism," *Philosophy and Phenomenological Research* 56:3 (1996): 567–81, §II, for a nice version of the argument. It conflicts with the suggestion that one's beliefs may be justified in the new evil demon case, where justification obviously does not require ruling out skeptical hypotheses. You might have guessed my response to these confusions about the nature of justification – they do not matter because we should quit thinking of justification as necessary for knowledge anyway.

3 Sensitivity

1 This is consistent with claiming that modalized process reliability generally requires responsiveness to the way the world is, a theme to which I will return in Chapters 4–6. It's just that in specific cases, like the Mary problem, there is a loophole, via double luck, in the process reliabilist account.

2 Robert Nozick, *Philosophical Explanations* (Cambridge, MA: Harvard University Press, 1981), 211; hereafter cited as Nozick (1981).

3 Nozick is aware (as I am) that sensitivity is not itself a relevant alternatives account (for it makes no mention of relevant alternatives). He does claim, however, that the subjunctive view can help to clarify which alternatives are relevant, and he even suggests that (3) may be taken as a criterion of relevance, in which case the relevant alternatives view "then becomes merely an alternate and longer

way of stating condition (3)" (Nozick 1981, 175). If only things were even that simple. We will soon see that which possible worlds are deemed relevant informs our interpretation of the subjunctive conditional!

4 Fred Dretske, "Conclusive reasons," *Australasian Journal of Philosophy* 49:1 (1971): 1–22, pp. 7–8; hereafter cited as Dretske (1971).

5 If this interpretation is feasible, it constitutes a basis for a reply to Goldman's Judy/Trudy case, described in his review of Nozick's book. Alvin Goldman, "Review of philosophical explanations," *Philosophical Review* 92:1 (1983): 81–8, p. 78; hereafter cited as Goldman (1983).

6 This assumes that I have the concepts to express the proposition that I am not a BIV. If content externalism is true, having these concepts requires that I have had some causal contact with actual brains, vats, and the like. See Putnam, *Reason, Truth and History* (Cambridge: Cambridge University Press, 1981), in particular "Brains in a vat."

7 The problem appears throughout the epistemology literature, and is originally formulated by Kyburg as a paradox because the odds against any one ticket lead one to believe, of each and every individual ticket, including one's own, that it will lose, even though one knows that one ticket will win. Henry Kyburg, *Probability and the Logic of Rational Belief* (Middletown, CT: Wesleyan University Press, 1961). The lottery problem is paradoxical in other ways, too. See John Hawthorne, *Knowledge and Lotteries* (Oxford: Clarendon Press, 2004); hereafter cited as Hawthorne (2004).

8 Alvin Plantinga, *Warrant and Proper Function* (New York: Oxford University Press, 1993), 199. Nozick considers a similar case (Nozick 1981, 190).

9 The example is from Keith Lehrer and Thomas Paxson, "Knowledge: undefeated justified true belief," *Journal of Philosophy* 66:8 (1969): 225–37, and it is discussed in Nozick (1981, 191–22).

10 When we come to the refinement according to which assessment of Nozick's conditions must be relativized to the method of belief formation actually used (§B) – which, surprisingly, comes *before* Nozick's discussion of this case – it will appear that S's belief does *not* violate condition (4). In my view, that's okay, first, because I reject condition (4), but more importantly for present purposes, because I think that it's not obvious that S does not know that Tom Grabit took the book.

11 Gilbert Harman, *Thought* (Princeton, NJ: Princeton University Press, 1973), Chapter 9.

12 But what if this is the only time the newspapers lied? It would follow that S's belief is formed by a reliable process. True, but then it becomes far less obvious to say that S does not know given the actual but unlikely subsequent denial that the dictator was shot. Lycan also seems to think that intuitions in the dictator case may very well disagree. William Lycan, "Evidence one does not possess," *Australasian Journal of Philosophy* 55:2 (1977): 114–26.

13 B.J. Garrett, "Nozick on knowledge," *Analysis* 43:4 (1983): 181–84, p. 183; hereafter cited as Garrett (1983).

14 Luper-Foy points out that Nozick's revised conditionals cannot handle "one-sided" methods – those that are capable of indicating that *p* is true, but not that *p* is false. The problem is that the antecedent of Nozick's conditional (3) is true only if S uses M to believe that *p* or not-*p*, but it is impossible, if M is one-sided, to believe that not-*p* on that basis, and thus S could only come to believe that *p* via M. If the antecedent can be satisfied only by believing that *p*, then the consequent, not believing that *p*, cannot be satisfied. Luper-Foy proposes the following revised construal of (3): If *p* were false, S would not believe that *p* via M. Here it is not required that S believe either *p* (in particular) or not-*p* (in particular) to satisfy the antecedent. Steven Luper-Foy, "The possibility of skepticism," in Steven Luper-Foy, ed. *The*

Possibility of Knowledge: Nozick and His Critics (Totowa, NJ: Rowman & Littlefield, 1987): 219–41, p. 225.

15 Jonathan Vogel, "Tracking, closure, and inductive knowledge," in Steven Luper-Foy, ed. *The Possibility of Knowledge: Nozick and His Critics* (Totawa, NJ: Rowman & Littlefield, 1987): 197–215, pp. 198–99; hereafter cited as Vogel (1987).

16 That methods are individuated from the inside does not entail that the method relevant to assessing sensitivity is the one the agent *believes* she is using. For example, the grandmother's belief that her grandson is well might in fact be based on the method: *If my grandson looks happy and rosy, believe that he is well.* She may wrongly think that her belief is based on some subtle difference in the way the grandson talks or on the testimony of other family members, but her actual method is a means of distinguishing, via the "final upshot in experience," the health of her grandson from the possibility that he is sick. The importance of fixing on the method actually used is discussed by Pritchard, *Epistemic Luck* (Oxford: Oxford University Press, 2005), 152ff.

Nozick's claim that the generality of methods "is set by differences the person would notice" seems to suggest a phenomenal interpretation of methods, which I think goes beyond the claim that two methods with the final upshot in experience are the same. Cases like the chicken sexer suggest that the phenomenal interpretation is false, for, by hypothesis, the chicken sexer is ignorant of the experiential cues she uses to determine the sex of a chick. Suppose that the chicken sexer in fact forms true beliefs on the basis of visual cues, but for some reason thinks that she does it by smell. The actual method is the relevant one, and those visual cues have a "final upshot in experience" – she *does* base her belief on what she sees – even though what she sees does not have a conscious or phenomenal impact on her. It is plausible to say that much of her method is sub-cognitive (and sub-personal). I therefore reject the phenomenal interpretation of methods, but retain Nozick's basic idea that two methods are identical if they have the same upshot in experience. Issues about the individuation of *experience* are therefore relevant, and I take them up briefly in note 19, below. They are not, however, directly relevant to the main problems afflicting Nozick's proposal for individuating methods that I discuss in the text, hence their relegation to an endnote.

17 See Robert Shope, "Cognitive abilities, conditionals, and knowledge: a response to Nozick," *Journal of Philosophy* 81:1 (1984): 29–48; hereafter cited as Shope (1984). Goldman (1983) and Garrett (1983) raise similar concerns, as do Graeme Forbes ("Nozick on scepticism," *The Philosophical Quarterly* 34:134 (1984): 43–52; hereafter cited as Forbes (1984)) and Luper-Foy ("The epistemic predicament: knowledge, Nozickian tracking, and scepticism," *Australasian Journal of Philosophy* 62:1 (1984): 26–49; hereafter cited as Luper-Foy (1984)).

18 Notice here that it's not enough to say that the *content* of the method is fixed by actual experience. S's method is believing it is Jesse James based on seeing certain facial features. If the mask had not slipped and S had used this method, she would not believe it is Jesse James, violating (4). Nozick has to claim, then, that application of the same method in counterfactual circumstances requires holding experience fixed in those circumstances, too, such that the method does not apply in nearby worlds where one's experience is different. But that seems to preclude the generality a method must have if it is to "play the appropriate role in subjunctives." To secure that role, one would think that methods must be applicable to different possible experiences. For example, if the animal Oscar sees were a hyena, then, applying his actual method, he would not believe it is a dog.

The failure of Nozick's proposal that actual experience must be held fixed is even more apparent when it is applied to condition (3). It makes little sense to say that, if it were not Jesse James, S would not believe it is, given her method, for, if actual experience must be held constant, S *would* believe it is Jesse James whenever she uses that method. The spirit of Nozick's response to the Jesse James case suggests that he would say simply that the method does not apply, and so S would not believe, in any case where the method is applicable, that it is Jesse James if it were not. But again, is it not desirable to individuate methods in a way that makes possible their application to different possible experiences such that, when applied, they are the basis upon which one is able to distinguish what is the case from close counter-possibilities?

Another way to put this point is to say that holding actual experience fixed for application of the method makes all methods one-sided – capable only of indicating that *p* is true, for in other situations the method is inapplicable (see note 14). Again, this undermines the generality of the applicability of methods. Suppose I believe that I have $197.32 in the bank based on reading my deposit receipt. We don't want to say that the method would not be applicable were the receipt to say something else, but that, if it were not $197.32, I would not believe that it is because it would give some other balance and I would read *it*.

We need to hold actual experience fixed when determining the content of the method, but not require that one have one's actual experience in counterfactual situations for application of the method in those worlds. This issue is, however, independent of the inconsistency charge, which arises because Nozick sometimes does not hold the method constant *at all*.

19 Luper-Foy (1984, 29) raises an interesting issue about method individuation, briefly discussed in note 16, namely, whether methods should be individuated by reference to the actual causes of belief or given a phenomenalist interpretation. He concludes, tentatively, the latter (at least for perceptual knowledge), which is consistent with Nozick's claim that methods that are the same "from the inside" are identical.

Williamson argues that methods should be individuated externalistically, which (I think) means by reference to external causes. Timothy Williamson, *Knowledge and its Limits* (New York: Oxford University Press, 2000); hereafter cited as Williamson (2000). Thus if one's method involves basing belief on one's experience, and one's experience is externalistically individuated, then one may very well use a different method in the case where one is a BIV than one does in the case where one is not a BIV. In this way, Williamson achieves some anti-skeptical consequences, since my belief that I am not a BIV, if true, is formed on the basis of a different method – on different experience – than it would be were I a BIV. Thus on the *method actually used*, it is not true that, if my belief that I am not a BIV were false, I would still believe it.

A crucial aspect of Williamson's defense of the claim that the methods in these cases are different hinges on his argument for the nontransparency of experience. (The argument is a kind of sorites, though Williamson argues that it is not thereby fallacious.) The fact that we cannot always distinguish our evidence implies that evidence is not something "internal" – in the sense of cognitively accessible – to the agent. It seems to me that we can accept the nontransparency (or anti-luminosity, as he calls it) argument without going so far as to say that experience is individuated externalistically, by reference to actual causes. (A method may thus be external in the sense that it is not cognitively accessible without also being external in the sense Williamson proposes.) Williamson's argument shows only that one's evidence (hence method) is not always cognitively available, which is something an externalist about knowledge should accept anyway.

Neither is it the case that accepting *content* externalism forces one to individuate experience externally. Consider again the chicken sexer. The chicken sexer knows nothing about the evidence, hence method, that she uses in determining the sex of a chick. Suppose the chicken sexer has recently been envatted. She therefore has the concepts *brain*, *vat*, *chicken*, *male*, *female*, and so on. When she forms the belief that *that* chick is a male, her thought token expressible by "that chick" fails to refer to a particular chick, and if methods are individuated by content, one could say that this is a different method than one used by a non-BIV chicken sexer. One might conclude that, once we accept content externalism and the denial of cognitive accessibility of methods, we need to individuate methods (and experiences) by reference to their actual causes. But now consider another of the envatted chicken sexer's beliefs, expressed by "I am not a BIV." On the assumption that she is recently envatted, this sentence means *I am not a BIV* even on a semantic externalist individuation. Surely it is not compulsory, in this case, to say that since the actual causes of her belief are electrodes and such, her method (as determined by content) is different than in the case where she truly believes that she is not a BIV. Again, it is not the case that she lacks the relevant concepts – *brain*, *vat*, and the like. As Williamson notes, "things appear to one in exactly the same way in the good [non-BIV] and bad [BIV] cases" (Williamson 2000, 165). Contra Williamson, I think this is a good reason for saying one's experience and evidence (hence method) are identical in both cases. Thus, for example, I do not know I am not a BIV because, if I were, I would still think *that I'm not a BIV*, based on my experience-cum-method. The arguments from anti-luminosity and content externalism do not require one to reject this otherwise deeply intuitive individuation of methods, according to which two method tokens are of an identical type if (1) their belief contents are identical; and (2) they involve the same capacity to distinguish objects, properties, facts, events, and the like. In subsequent chapters, the capacity to distinguish will be a central theme. In any case, Williamson's response to the skeptic is troublesome. I can know I'm not a BIV because I have positive evidence to that effect, but my positive evidence is indistinguishable (from my perspective) from the negative evidence I would have were I a BIV. My counterproposal, to put it bluntly, is that experience need not be available phenomenally to the agent to be distinguishing, but it must involve a distinguishing capacity to be a basis for knowledge.

What, then, of Nozick's claim that methods are determined by what one would notice? *When* one would notice differences in methods, those methods are in fact different, though there are cases where the method is cognitively inaccessible, perhaps because sub-cognitive. But even then, the method actually used is one that involves a distinguishing capacity, for instance the ability to tell male from female chicks by sight, even if the chicken sexer does not know that this is her method. I want to thank Sandy Goldberg for urging me to pay closer attention to the issues in this note.

20 For present purposes, "experience" includes, among other things, evidence obtained through testimony, and is thus a more general notion than direct perceptual experience. On the other hand, it cannot plausibly be the primary basis upon which one normally gains a priori knowledge, where one's method involves a specific bit of reasoning or perhaps even intuition. Thus not every method is an experiential one.

21 Edward Craig, "Nozick and the sceptic: the thumbnail version," *Analysis* 49:4 (1989): 161–2.

22 Anthony Brueckner, "Unfair to Nozick," *Analysis* 51:1 (1991): 61–4.

23 Alan Goldman, "Nozick on knowledge: finding the right connection," in Steven Luper-Foy, ed. *The Possibility of Knowledge: Nozick and His Critics* (Totowa, NJ: Rowman & Littlefield, 1987), 182–96, p. 184.

24 Shope seems to agree that a global reliability condition can help in the Haveit case, but he is concerned that the process must be construed perhaps too *generally* to work. "The process will turn out to be unreliable if characterized at a higher level of generality, for example, as forming a belief guided by evidence that has been unsuspectedly fabricated. But a reliability theorist needs a rationale

for ascending to that level of generality." Robert Shope, "Conditions and analyses of knowing," in Paul Moser, ed. *The Oxford Handbook of Epistemology* (Oxford: Oxford University Press, 2002), 25–70, p. 38. The context of these remarks, however, is a discussion of subjunctives, that is, of local reliabilism, where, we have seen, the method actually used must be fairly *narrowly* individuated, including specific content. My proposal is (or will be) that we *add* an independent global reliability condition, and that reliable processes are naturally individuated more widely – at least content-neutrally.

25 Ernest Sosa, "How to defeat opposition to Moore," *Philosophical Perspectives* 13 (1999): 141–53, p. 145; hereafter cited as Sosa (1999).

26 Building receptivity both to one's actual positive evidence and to possible counterevidence into the method provides a basis for a partial reply to the worry that Nozick is committed to illicit "backtracking" counterfactuals (Vogel 1987, §V and Hawthorne 2004, 13), such as: If the sun had not come up, things (including the belief-forming method used) would have been very different beforehand. Backtracking counterfactuals are (putatively) independently problematic, and, when Nozick appeals to them to solve the sunrise case, he is being inconsistent in his application of methods, for he is appealing to a different method of basing belief were the sun not to rise – there would have been evidence that the sun would not come up – from the normal one used when the sun is going to rise. I am claiming, however, that the method used is *not* different, and so one need not claim that it would have been.

27 How would an agent even know that the outcome is likely? If it were false that it is likely and she is not attuned to counterevidence, then she would believe it is likely anyway. The problem with this objection is to think that an event is unlikely just because in the case in question it in fact will not happen. If a meteor were about to smash into the earth disrupting its rotation, there is a sense in which it is false that it is likely that the sun will "rise." But that is like saying that, because in fact the top card is an ace, it is not (objectively) unlikely that it is an ace, or that, because some particular person will win the lottery, it is not unlikely that she will. Similarly, the fact that a meteor is about to strike the earth does not make it likely that the sun will not rise, because the meteor event is itself, while actual (let us say), unlikely. So the antecedent of the conditional – if it were false that it is likely – is not satisfied.

Brian Cooney has generously offered this suggestion:

> To know that an event is likely is to know that the proposition asserting that the event will happen is likely to be true. This would remain the case when something unknown to the speaker will make the sentence false *this time*. It is always the case that such propositions are likely to be true if well grounded in induction, even though it is recognized that they might be false in a few cases. One may even be in possession of the precise statistical likelihood.

28 Jonathan Vogel, "Reliabilism leveled," *The Journal of Philosophy* 97:11 (2000): 602–23; hereafter cited as Vogel (2000). The following explication of Vogel's argument and my reply to it are taken, for the most part, from my paper, "Is counterfactual reliabilism compatible with higher-level knowledge?" *Dialectica* 60:1 (2006): 79–84.

29 I have recently come across a paper by Joe Salerno that responds to Vogel in much the same way I do here and in the paper cited above. Joseph Salerno, "Truth-tracking and the problem of reflective self-knowledge," in Joseph Keim-Campbell and Michael O'Rourke, eds. *Topics in Contemporary Philosophy* 5: *Knowledge and Scepticism* (Cambridge, MA: MIT Press, in press).

30 That is, you believe that the first-order *proposition* believed is true. For both ease of exposition and consistency with some of Vogel's locutions (for example, "Consider the nearest possible world W in which your higher-level belief is false"), I will often just use the shorter "believe that one's belief is true."

31 This assumes that introspection failure is not the issue, that is, that you would not believe that you believe that *p* unless you believe that *p* (whether your belief that *p* is true or not).

4 Safety

1 Ernest Sosa, "How to defeat opposition to Moore," *Philosophical Perspectives* 13 (1999): 141–53, pp. 145–46; hereafter cited as Sosa (1999).

2 The third formulation might be construed as approximating sensitivity, since a natural reading of "not easily would S believe that p without it being the case that p" is: *if p were false, S would not believe that p.* In aiming to distinguish safety from sensitivity, Sosa clearly does not intend this reading. On this point, see also Sherrilyn Roush, *Tracking Truth* (Oxford: Oxford University Press, 2005), 123; hereafter cited as Roush (2005).

3 Duncan Pritchard, "Resurrecting the Moorean response to the sceptic," *International Journal of Philosophical Studies* 10:3 (2002a): 283–307, p. 295; hereafter cited as Pritchard (2002a, "Moorean response").

4 Duncan Pritchard, "Virtue epistemology and epistemic luck," *Metaphilosophy* 34:1/2 (2003): 106–30, p. 118; hereafter cited as Pritchard (2003).

5 The idea is that the realm of safety that applies to first-level beliefs also applies to higher-level beliefs about them. I have used this basic insight to refute the incompatibility of sensitivity and higher-level knowledge in the previous chapter, though of course the story is somewhat more complicated.

6 See G.E. Moore, *Philosophical Papers* (London: George Allen & Unwin, 1959), in particular "A defence of common sense" and "Proof of an external world."

7 Hawthorne presents a similar problem for (neo-)Mooreans. John Hawthorne, *Knowledge and Lotteries* (Oxford: Clarendon Press, 2004), §3.7.

8 John Greco, "Virtue and luck, epistemic and otherwise," *Metaphilosophy* 34:3 (2003a): 353–66, p. 359; hereafter cited as Greco (2003a, "Virtue and luck").

9 Linda Zagzebski, *Virtues of the Mind* (Cambridge: Cambridge University Press, 1996), 286; hereafter cited as Zagzebski (1996).

10 We can also reverse the reasoning: If the intuition that I (in fact) know that the bag will soon be in the basement is stable, it follows that there are not (or cannot be) close worlds where the bag gets caught. My thanks to Duncan Pritchard for putting it this way. His book includes a discussion of this case. *Epistemic Luck* (Oxford: Oxford University Press, 2005), 164f.

11 Juan Comesaña, "Unsafe knowledge," *Synthese* 146 (2005): 395–404, p. 397.

12 This is another case that suggests the pure subjunctive reading of the sensitivity conditional, since there is indeed a nearby world where I form a false belief about the location of the party, but it is not among the closest possible worlds.

13 Sosa's virtue theory is committed to this point about safety, and indeed in places he suggests that a belief is not safe in the relevant sense just by satisfying the safety principle as described in the text: "what is required for a belief to be safe is not just that it would be held only if true, but rather that it be based on a reliable indication." Ernest Sosa, "Skepticism and contextualism," *Philosophical Issues* 10 (2000): 1–18, note 10.

14 A slightly longer version of Pritchard's argument and my reply constitutes my paper, "Reliabilism and safety," *Metaphilosophy* 37:5 (2006): 691–704.

15 That's not exactly right, since Pritchard entertains the possibility that a responsibilist virtue theory may be required to eliminate *reflective* luck, but neither

Pritchard nor I think that the elimination of reflective luck is necessary for knowing. Recall that a belief is reflectively lucky just in case it is true in the actual world but false in nearby worlds consistent with what one knows by reflectively accessible means. Reflective luck cases such as the chicken sexer suggest to some that there must be an internalist (and typically responsibilist) justificatory condition on knowledge. But as I hope is obvious by now, I think it may be correct to attribute knowledge to the chicken sexer, provided that the chicken sexing ability is not merely "accidental," where "accidental" means *produces true beliefs only in the actual world.* A very general argument for externalism can be found in my "Knowing and possessing knowledge," *American Philosophical Quarterly* 41:1 (2004): 21–36.

16 See John Greco, *Putting Skeptics in Their Place* (Cambridge: Cambridge University Press, 2000), 175–6; hereafter cited as Greco (2000). I use the general term "knowledge-precluding" to describe the benevolent demon, rather than either "veritic" or "process" luck for reasons described in Chapter 2, note 19. The process produces many false beliefs in nearby worlds, which suggests process luck, but one could say that there are many worlds where René's belief, for example that 17 is due to come up next on the roulette wheel, based on his fallacious reasoning, might easily have been false, and so constitutes veritic luck. Later in this chapter I will use the phenomenon of process luck to undermine the safety-based account. In the examples produced, it will be clear that veritic luck, as Pritchard defines it, is not the central issue.

17 Alvin Plantinga, *Warrant and Proper Function* (New York: Oxford University Press, 1993), 199.

18 See, for example, Greco (2000) (Chapters 7 and 8) and "Virtue and luck," and Sosa, *Knowledge in Perspective* (Cambridge: Cambridge University Press, 1991).

19 The example is a little peculiar, but so is John! The point is that any such method is likely to lead to false beliefs. If John believes he's not in the Matrix because he hates Keanu Reeves, then perhaps in a world where he applies his method but Edward Norton stars in the movie, he is susceptible to believing what he sees in the film.

20 The idea that John might use some other method raises the following possible reply: John's belief is safe *but constitutes knowledge anyway*, because the method attributed to John is simply not the salient one. John's belief is overdetermined by methods – formed on the basis of the *emotional reaction to the star* method, and on the method through which John notices that his ordinary empirical beliefs are inconsistent with the Matrix hypothesis and thereby comes to believe that he is not in the Matrix. Recall the discussion of overdetermination of methods in the previous chapter, based on the faithful father example. There we said that the father knows only if the method based on the jury's conviction "outweighs" his faith, that is, if in the possible world where the son is guilty, the father would accept (the method based on) the jury's verdict and give up (the method based) on faith. We concluded, more generally, that one knows that *p* only if, when one's belief that *p* is overdetermined by methods, were the proposition believed false, the method that outweighs all others would lead one not to believe it. If we apply this lesson to safety, it is plausible to say that if the *emotional reaction to the star* method and the inference from ordinary beliefs methods were to deliver incompatible verdicts, John would (if he were not paranoid) base his belief about whether he is in the Matrix on the latter, and that method would produce a safe belief. In what follows in the text I set aside this reply. We can stipulate, by appeal to his general paranoia, that John's only method is the hated actor one.

21 The quotation goes on to say: "(i.e., where ϕ is true, she believes it; where ϕ is false, she does not believe it)." But this is a misprint. It should say: (i.e., where she believes it, ϕ is true; where she does not believe it, ϕ is false). Thanks to Duncan Pritchard for pointing this out to me.

22 Though the expression "deeply contingent" does not appear in his recent book, Pritchard has made this claim in conversation.

23 Donald Davidson, "Mental events," in *Essays on Actions and Events* (Oxford: Clarendon Press, 1980), 207–25.

24 Alvin Goldman, "Discrimination and perceptual knowledge," *Journal of Philosophy* 73:20 (1976): 771–91, p. 771.

25 Colin McGinn, "The concept of knowledge," in Peter French, Theodore Uehling, and Howard Wettstein, eds. *Midwest Studies in Philosophy* 9, *Causation and Causal Theories* (1984), 529–54, p. 530 (emphasis in original).

26 Christopher Hill, "Process reliabilism and Cartesian scepticism," *Philosophy and Phenomenological Research* 56:3 (1996): 567–81.

27 Surely intuitions may diverge here. One could deny that a discriminating capacity is necessary for knowledge. Or one could say that reliability does not entail the ability to discriminate. My overall strategy in this chapter is to militate against these moves. First, I want to show that some version of process (or agent or faculty) reliabilism is a necessary feature of a theory of knowledge, which renders local versions of reliabilism insufficient, in particular safety, which does not require discrimination. Here (§E) I aim to show how genuine reliability requires a discriminative capacity. Putting the points together, it follows that one cannot know that *p* unless one can discriminate it from possibilities – not necessarily all of them, just the closest ones – where not-*p*. One need not *accept* all of this; my arguments simply aim to make these ideas compelling.

28 I want to thank Duncan Pritchard for urging me not to overlook this position.

29 Jonathan Schaffer, "Contrastive knowledge," in Tamar Gendler and John Hawthorne, eds. *Oxford Studies in Epistemology: 1* (Oxford: Oxford University Press, 2005), 235–71, p. 235; cited next as Schaffer (2005), and subsequently just with the page number.

30 The discrimination requirement is implicit here. In a previous draft of this paper, Shaffer characterized knowledge thus: "s knows that *p* rather than *Q* if and only if *s* decisively discriminates *p* from *Q*."

31 Stalnaker claims that the syntactic arguments given by Schaffer for ternary rather than binary propositional structure are not compelling, but in that context his concern is to show that contrastivism is not materially different from contextualism. Robert Stalnaker, "Comments on 'From contextualism to contrastivism'," *Philosophical Studies* 119 (2004): 105–17. Because I also reject contextualism, I will not enlist Stalnaker's arguments in my project.

32 Schaffer notes that some inquiry-guiding questions are based on false presuppositions, from which it simply follows that none of the answers are true. But the goldfinch or raven case need not be one of false presupposition. It may be true that either it is a goldfinch or a raven. Still, if for all I know it is a canary, then my ability to rule out its being a raven does not yield knowledge that it is a *goldfinch*, rather than a raven.

33 This could sound question-begging against the contrastivist. Schaffer points out that one's ability to answer the question of whether it is a goldfinch or a raven is inexplicable if knowledge ascriptions are construed as having binary propositional form (244). Absolutely right. But what contrastive knowledge does one actually have here? Not that it is a goldfinch rather than a raven, but that it is some kind of yellow bird rather than a raven, and there's your progress.

34 And it is not entirely clear what other form of genuine responsiveness there could be.

5 Knowledge: Reliably formed sensitive true belief

1 I have said very little about the nature of belief, mostly because it is not the main concern of this book. L.J. Cohen argues that we need to distinguish belief from

acceptance, and claims that different conceptions of knowledge derive, respectively, from them. L. Jonathan Cohen, *An Essay on Belief and Acceptance* (Oxford: Oxford University Press, 1992). If one were to think of belief as Cohen thinks of acceptance, roughly, as voluntary, then one might deny that belief is necessary for knowledge. Whatever the right understanding of this condition, it has to involve something like S taking it to be true that *p*, and I think 'believes that' suffices for our purposes.

2 As noted in the introduction (§C.2), the second and fourth conditions entail the first, since if S in fact believes that *p* and S's belief is sensitive, *p* must be true, for if it were false, S would not believe it. The closest world where S sensitively believes that *p* is the actual world, so *p* must be true in the actual world. Strictly speaking, then, condition (1) is redundant. Though it should be clear from Chapter 4, I want to emphasize that sensitivity is not equivalent to safety – in nearby worlds where S believes that *p*, *p* is true – by contraposition, because a safe belief must be true only in nearby worlds, whereas sensitivity sometimes requires that one not hold a belief, for instance, that I am not a BIV, in distant worlds (but still the nearest ones) where it is false.

3 Alvin Goldman, *Epistemology and Cognition* (Cambridge, MA: Harvard University Press, 1986), 93; hereafter cited as Goldman (1986). See also Goldman "A priori warrant and naturalistic epistemology," in *Pathways to Knowledge* (New York: Oxford University Press, 2002), 24–50, p. 38.

4 Again, I must stress that here I am imagining that we have settled the question of the *distance* of the worlds through which one's belief-forming processes must be truth-conducive, and I am asking: In how many of the worlds *within* that distance must the process must be truth-conducive? There may be no precise measure of distance. Rather, the process must be truth-conducive throughout the nearest possible worlds, and it will be more reliable still if truth-conducive in more distant worlds, hence that knowledge will be more secure. What follows in the text is a defense of the claim that distant worlds are not relevant to judgments of reliability, with no pretensions to finding an exact cut-off line between the nearby and distant worlds relevant to those judgments.

5 This particular example might cause confusion about the sense of "narrow" at work. By "narrow" I mean refined or specific, as opposed to its sense in "narrow content," that is, mental content individuated by reference solely to states of the individual (which also tends to be more specific than wide, or referential, content). In the example, states of the individual are changed due to the work of the demon, and this may suggest the latter kind of narrowness, but I hope that it is clear from discussions of process and method individuation throughout this book that this is not primarily what I mean by narrowness as it applies to belief-forming processes. See also Chapter 3, note 16.

6 Duncan Pritchard, "Virtue epistemology and epistemic luck," *Metaphilosophy* 34:1/2 (2003): 106–30, §§1 and 3.

7 Alvin Plantinga, *Warrant and Proper Function* (New York: Oxford University Press, 1993).

8 John Greco, "Virtue and luck, epistemic and otherwise," *Metaphilosophy* 34:3 (2003a): 353–66, p. 357; hereafter cited as Greco (2003a, "Virtue and luck").

9 See, for example, the following works by Linda Zagzebski: *Virtues of the Mind* (Cambridge: Cambridge University Press, 1996); "The search for the source of epistemic good," *Metaphilosophy* 34:1/2 (2003a): 12–28; "Intellectual motivation and the good of truth," in Michael DePaul and Linda Zagzebski, eds. *Intellectual Virtue* (Oxford: Oxford University Press, 2003b), 135–54.

10 Responsibilism invites the question of whether agents even have doxastic control, which I am happy to be permitted, as an externalist, to ignore. Alston argues that thinkers don't have voluntary control over beliefs, and so deontological

concepts are not central to knowledge because they imply voluntarism. William Alston, "The deontological conception of epistemic justification," *Philosophical Perspectives* 2 (1988): 257–99. Heil argues that thinkers have *indirect* control over beliefs. John Heil, "Doxastic agency," *Philosophical Studies* 43:3 (1983): 355–64. Feldman argues that epistemic obligation does not entail voluntarism, which implies that Alston's view, even if right, is not entirely relevant to the question of whether obligations attach to what one believes. Richard Feldman, "Epistemic obligations," *Philosophical Perspectives* 2 (1988): 235–56. Steup argues that in fact thinkers do have control over their beliefs. Matthias Steup, "A defense of internalism," in Louis Pojman, ed. *The Theory of Knowledge*, 3rd edition (Belmont, CA: Wadsworth, 2003): 310–21. As I say, it's nice to be able to avoid this controversy.

11 See John Greco, *Putting Skeptics in Their Place* (Cambridge: Cambridge University Press, 2000), Chapter 8; hereafter cited as Greco (2000). See also Greco (2003a, "Virtue and luck," 363).

12 John Rawls, *A Theory of Justice* (Cambridge, MA: Harvard University Press, 1971), 120; hereafter cited as Rawls (1971).

13 *Very* briefly, the worry goes like this. Credit *may* imply responsibility. Responsibility may and usually does imply some element of direct control. There is no direct control of those elements of the belief-truth link external to – not recognizable on reflection by – the agent. Therefore, justification must be internalist. It is just a worry. Any step in the argument might be challenged, but it is prima facie plausible.

14 I just implied that strength of epistemic position is a function of sensitivity. DeRose characterizes strength of epistemic position differently, which has consequences for how he incorporates sensitivity into his contextualist theory. Keith DeRose, "Solving the skeptical problem," in Keith DeRose and Ted Warfield, eds. *Skepticism: A Contemporary Reader* (Oxford: Oxford University Press, 1999), 183–219. See also Chapter 6, below.

15 See Robert Nozick, *Philosophical Explanations* (Cambridge, MA: Harvard University Press, 1981), 174–5; hereafter cited as Nozick (1981).

16 Aristotle, "Nichomachean ethics," in J.L. Ackrill, ed. *A New Aristotle Reader* (Princeton: Princeton University Press, 1987), 1094b12–14.

17 David Lewis, *Counterfactuals* (Oxford: Blackwell, 1973), §1.7.

18 B.J. Garrett, "Nozick on knowledge," *Analysis* 43:4 (1983): 181–84, p. 183.

19 "Aha!" you say. "But would it produce a *significant ratio* of false beliefs in nearby worlds, enough to undermine the "high ratio" requirement of the reliability condition?" I think so. It is plausible that different kinds of belief-forming processes will have different standards for what constitutes a high ratio. Given the exactness of mathematics, a process that predictably results in some false beliefs – not merely due to glitches like forgetting to carry or inattention – falls short of the high ratio requirement in its domain. So even if the "no counterexamples" method used by our hypothesized mathematician is statistically truth-conducive (which is not something I am granting, but only supposing), it fails to be sufficiently truth-conducive for its domain.

20 Steven Luper-Foy, "The epistemic predicament: knowledge, Nozickian tracking, and scepticism," *Australasian Journal of Philosophy* 62:1 (1984): 26–49, p. 31.

21 The case is from Alan Goldman, "Nozick on knowledge: finding the right connection," in Steven Luper-Foy, ed. *The Possibility of Knowledge: Nozick and His Critics* (Totowa, NJ: Rowman & Littlefield, 1987), 182–96, p. 184.

22 Richard Fumerton, "Nozick's epistemology," in Steven Luper-Foy, ed. *The Possibility of Knowledge: Nozick and His Critics* (Totowa, NJ: Rowman & Littlefield, 1987), 163–81, pp. 170–71. Other similar cases are discussed in Fumerton, *Metaepistemology and Skepticism* (Lanham, MD: Rowman & Littlefield, 1995); see especially Chapter 4.

23 A similar but perhaps more troubling example: I do not know both that I am not in a vat and that I am not a brain in a vat, even though I know the former and that it entails the latter. I do not know that the conjunction is true because I do not know that the second conjunct is true. This, again, is perhaps the most counterintuitive implication of the theory defended here, namely, the denial of closure. And again, this is a topic I take up in the final chapter.

24 As it did an anonymous reader, to whom I am indebted for pointing this out.

25 A further indication that my proposal here is not ad hoc is that it is consistent with my independently motivated remarks in §A.1 about the possibility that a demon can be part of a reliable process. Suppose I can distinguish chords, but only when I do not have a migraine. If the demon steps in and relieves my migraine, and he does this reliably, then his actions constitute part of a reliable process (though surely I deserve only "partial" credit here – none for the demon's help). The listening sub-process by which I distinguish chords is not separable from the (demon) basis upon which it is successfully implemented. Similarly, the $method_w$ by which the naturally gifted math student achieves correct calculations is not separable from the basis upon which she chooses the algorithm.

26 I doubt that Greco would accept the idea that knowledge requires understanding of the processes used, given the general externalist thrust of his thinking, and so I mean it quite literally when I say that the case of the careless math student is (merely) "apt to mislead." It is certainly not the case that Greco argues explicitly along the lines set out and rejected in the previous paragraph.

27 Earl Conee and Richard Feldman, "The generality problem for reliabilism," in Louis Pojman, ed. *The Theory of Knowledge*, 3rd edition (Belmont, CA: Wadsworth, 2003), 296–309, p. 297; hereafter cited as Conee and Feldman (2003).

28 They also claim that the solution must be in the spirit of reliabilism, which means that it must not be couched in evaluative terms. I leave it to the reader to decide whether the solution I shall offer violates this criterion. Obviously, I do not think so.

29 This is essentially a combination of distinct ideas put forth by Goldman, but, so far as I know, he does not put them together. On content-neutrality, see Goldman, "What is justified belief?" in George Pappas, ed. *Justification and Knowledge* (Dordrecht: D. Reidel, 1979), 1–23, p. 12; hereafter cited as Goldman (1979). On his claim about the narrowest causally operative type, see Goldman (1986, 50–1).

Why not instead try to capture the requisite fineness of processes by appeal to "normal" or "good" conditions? The problem with this suggestion is that "normal" or "good" are themselves undefined notions, and if all that is meant is *conditions where one's belief-forming processes are truth-conducive*, one would wonder why we bothered to introduce the notions of normal or good into the account; they would be doing absolutely no work. Worse still, it is not clear that reference to normal or good conditions can capture the way in which process types can and should be individuated by reference to a given agent's particular cognitive abilities, accounting for which, I shall argue, is important.

30 I will not discuss proposed solutions that have no bearing on the present proposal. For instance, Schmitt claims that we do not need a precise criterion for process types; rather, we can let our intuitions about which processes are relevant do the work. Frederick Schmitt, *Knowledge and Belief* (New York: Routledge, 1992), Chapter 6. Heller also argues that we need no necessary and sufficient conditions, and that true statements about the relevant operative process types are determined by contextual factors. Mark Heller, "The simple solution to the problem of generality," *Noûs* 29:4 (1995): 501–15.

31 This strategy is drawn from William Alston, "How to think about reliability," *Philosophical Topics* 23:1 (1995): 1–29; hereafter cited as Alston (1995). A new version appears in Alston, *Beyond Justification* (Ithaca: Cornell University Press, 2005).

32 Richard Feldman, "Reliability and justification," *The Monist* 68:2 (1985): 159–74, p. 161.

33 Maybe I should not assert this so straightforwardly. My solution to the generality problem requires that process individuation abstracts from the *specific* content of the particular belief in question, which will make room for some processes that are perhaps even less content-neutral than those Goldman has in mind.

34 Robert Shope, "Conditions and analyses of knowing," in Paul Moser, ed. *The Oxford Handbook of Epistemology* (Oxford: Oxford University Press, 2002), 25–70, p. 38.

35 "But what if Haveit, for whatever reason, aims to dupe only in this one instance? Suppose that everything else he says is true and not misleading. Then isn't his testimony reliable?" One thing to say here is that my belief-forming process is unreliable in the most fundamental sense, in that I cannot distinguish true from false testimony. But I acknowledge that it is hard to find *any* individuation of processes or methods$_w$ that conforms to Haveit's testimony being unreliable in this case, and that the example is so constructed that appeal to the method$_n$ will not help either. Perhaps a more promising thing to say is that I *do know* that someone in the office owns a Ford. Haveit seems hell-bent on making sure that I have a true belief, and he is (by hypothesis) impeccably reliable! This, like the vase-tracking hologram case, is an extremely cleverly crafted case of double luck. I admitted in the introduction (§C.2) that it will always be possible to construct some such cases for any theory, but added that to truly undermine the theory of knowledge offered herein, the cases will be controversial, as these are.

How damaging is this? Well, notice again just how precisely the examples must be engineered. There must be lots of bad luck in that one's process is reliable, one's belief is sensitive, and yet the reliable process produces false grounds for belief, and there must be extremely good luck in that, in *just this kind of odd scenario*, the belief is in fact true. But is it really luck that the belief is true any more? Haveit is reliable and aims to make me have a true belief. (Again, if Haveit's testimony does not generally produce true beliefs, the processes are not reliable.) When so cleverly constructed, perhaps it is no longer clear that these are Gettier cases! In such extraordinary circumstances, knowledge could be based on (one-off cases of) misleading evidence. Let me add, though, that the tendency of the original Haveit case suggests that it is best described in terms of Haveit's testimony being unreliable, which is the diagnosis I give in the text.

6 Closure and the value problem

1 Riggs, "Reliability and the value of knowledge," *Philosophy and Phenomenological Research* 64:1 (2002): 79–96; hereafter cited as Riggs (2002). Though Riggs does not exploit the notion of a strong epistemic position, he locates the value of reliably formed beliefs not in the beliefs themselves, but in the reliable mechanisms that produce them. This could be seen as skirting the main issue, namely, explaining how a reliably formed true belief is better than a mere true belief, but that issue hinges on a contentious premise, namely, *that* a reliably formed true belief is better than a mere true belief. Surely when one knows, there is a source of value *somewhere* that is absent in cases of mere true believing, but it does not follow that the value is in the belief itself.

2 Another motivation for contextualism, expressed in Unger's influential *Philosophical Relativity*, is that the search for objective, invariantist (non-contextualist) answers to skeptical and other problems leads to debates that "appear endless. And little solid progress toward a solution as satisfactory as it is definite ever seems made." Peter Unger, *Philosophical Relativity* (Oxford: Oxford University Press, 2002), 4. I believe, however, that the reliably formed sensitive true belief theory *does* represent philosophical progress. It offers resources to answer the

skeptic, answers to our most stable intuitions about cases of knowledge, and (with its reliabilism) provides a research program for the integration of cognitive science and epistemology. Its primary alleged weakness is its rejection of closure.

3 And thus *would* know, assuming closure is true, and if I made the inference.

4 Peter Klein argues that versions of closure that do not admit of counterexamples are useless to the skeptic because the premises (especially P1) of the skeptic's closure-based argument need support, and any argument employed to that end already entails the conclusion, for instance, that I do not know that O, which renders the premises of the closure-based argument superfluous; hence the argument "*virtually* begs the question." Peter Klein, "Skepticism and closure: why the evil genius argument fails," *Philosophical Topics* 23:1 (1995): 213–36, p. 226. Klein thus sees a possibility for avoiding skepticism while accepting internalism and closure. See Pritchard, *Epistemic Luck* ((Oxford: Oxford University Press, 2005), 45–6; hereafter cited as Pritchard (2005)) for a critical discussion of Klein's strategy, which hinges on the claim that one has more evidence for the denials of skeptical propositions than one has for the ordinary propositions that entail the denials, the additional evidence including the ordinary proposition and the inference from it to the anti-skeptical proposition.

5 Stewart Cohen, "Contextualism and skepticism," *Philosophical Issues* 10 (2000): 94–107, p. 100; hereafter cited as Cohen (2000).

6 Keith DeRose, "Solving the skeptical problem," in Keith DeRose and Ted Warfield, eds. *Skepticism: A Contemporary Reader* (Oxford: Oxford University Press, 1999), 183–219, p. 184; hereafter cited as DeRose (1999).

7 Ram Neta, "Contextualism and the problem of the external world," *Philosophy and Phenomenological Research* 66:1 (2003): 1–31, p. 21.

8 Williams's inferential contextualism differs from, and is more radical than, the kind of semantic contextualism discussed here in, among other things, holding that there is no context-independent way of understanding how context can determine the standards relevant to assessing the epistemic status of a belief. It seems to follow that there is no univocal notion of "knowledge" about which to theorize. Michael Williams, *Unnatural Doubts: Epistemological Realism and the Basis of Scepticism* (Oxford: Blackwell, 1991). Contrasting this with, for example, Kornblith's claim that knowledge is a natural kind reveals the great distance between philosophers in their basic epistemological beliefs. Hilary Kornblith, *Knowledge and its Place in Nature* (Oxford: Clarendon Press, 2002).

9 Thus if we take seriously the contextualist's concern that any anti-skeptical epistemology must account for the worrisome nature of skeptical hypotheses, it is difficult to see how the safety theorist answers it, since safety seems modeled precisely just to make room for knowledge of the falsity of skeptical hypotheses and to maintain closure, without explaining how we achieve such knowledge, hence without giving proper weight to skeptical hypotheses.

10 See, for instance, Crispin Wright, "Cogency and question-begging: some reflections on McKinsey's paradox and Putnam's proof," *Philosophical Issues* 10 (2000): 140–63, and "Some reflections on the acquisition of warrant by inference," in Susana Nuccetelli, ed. *New Essays on Semantic Externalism and Self-Knowledge* (Cambridge, MA: MIT Press, 2003), 57–77; hereafter cited as Wright (2003). See also Martin Davies, "Externalism and armchair knowledge," in Paul Boghossian and Christopher Peacocke, eds. *New Essays on the A Priori* (Oxford: Oxford University Press, 2000), 384–414, and "The problem of armchair knowledge," in Susana Nuccetelli, ed. *New Essays on Semantic Externalism and Self-Knowledge* (Cambridge, MA: MIT Press, 2003), 23–55.

11 Jessica Brown, "The reductio argument and transmission of warrant," in Susana Nuccetelli, ed. *New Essays on Semantic Externalism and Self-Knowledge* (Cambridge, MA: MIT Press, 2003), 117–30, p. 121.

12 It is often claimed that sensitivity theorists have to deny closure only to make the theory come out right, but this is somewhat misleading. In a footnote to his discussion of Nozick's subjunctive conditional theory, Feldman writes: "That's not quite fair. Nozick doesn't just reject closure. Instead, he gives an analysis of knowledge that implies that closure is false. And then he defends the denial of closure on the basis of the analysis." Richard Feldman, "Contextualism and skepticism," *Philosophical Perspectives* 13 (1999): 91–114, note 11; hereafter cited as Feldman (1999). That is, Nozick would claim that the theory stands on its own, and the denial of closure is a consequence. Perhaps it is the denial of closure that has motivated philosophers to conjure counterexamples to Nozick's subjunctive conditional conditions. In any case, I am trying to *explain* the failure of closure within the framework of an independently motivated reliabilism.

13 Though I have not discussed it here, Hill argues that skeptical arguments held sway for so long because philosophers did not see externalism (generally) as an option for the belief-truth link, and this perhaps because they did not have the theoretical resources for distinguishing subjective and objective probability, the latter being particularly relevant to externalist accounts of "evidential" support. Christopher Hill, "Process reliabilism and Cartesian scepticism," *Philosophy and Phenomenological Research* 56:3 (1996): 567–81. This may be a plausible addition to a fuller account of the appeal of skeptical arguments that turn out not to work once we embrace reliabilism.

14 It is sometimes claimed that Dretske's denial of closure hinges on his accepting an implicitly internalist view of justification because Dretske claims that one has no *available evidence* to rule out such skeptical hypotheses, and that is why one does not know. (See, for example, Pritchard, "Radical scepticism, epistemological externalism, and closure," *Theoria* 68:2 (2002c): 129–61, and his 2005, Chapter 2.) My approach does not suffer this particular possible defect. The reason one does not know that not-SH is simply that one has no reliable way of establishing an *independent* warrant, internalistically or externalistically construed, for that belief. Violation of sensitivity bears this out, so with his appeal to sensitivity I think Dretske could (and should) have said this, too. In fact, he comes very close to saying just this: "Perception, our most important way of telling what is happening in the external world, doesn't tell us that there is an external world." Dretske, "Externalism and self-knowledge," in Susana Nuccetelli, ed. *New Essays on Semantic Externalism and Self-Knowledge* (Cambridge, MA: MIT Press, 2003), 131–42, p. 138. It is not just that we have no evidence for not-SH, but that our mechanisms for knowing ordinary propositions are useless for such propositions.

15 This desideratum may sound a little odd. Of course, we know *that* our ordinary beliefs would be false if skeptical hypotheses were true. But compelling skeptical scenarios should also *explain* our lack of knowledge – should explain how knowledge is undermined without merely *saying* that it would be because our beliefs would be false.

16 Herein also is a link between philosophical skepticism and everyday skepticism. Authoritarian governments (and, alas, others too) exploit our reliable cognitive mechanisms by censoring newspapers and getting falsities into print and on the air. Those normally reliable mechanisms – perception, reliance on testimony, and the like – are in many cases insufficient to put us in a position to know whether the reliably received misinformation is true. Magicians and liars also prey on the reliability of our cognitive processes. They trust that we'll see and hear *correctly* but misleadingly what they show and tell us, and so use our reliable belief-forming processes against us, rendering them useless. I am not claiming that the culprit in all compelling skeptical scenarios acts maliciously or even intentionally (or, for that matter, that the culprit is always an agent), but only that the scenarios offer an account of how our reliability is undermined.

17 Stewart Cohen, "Contextualism, skepticism, and the structure of reasons," *Philosophical Perspectives* 13 (1999): 57–89; see also Cohen (2000).

18 Worse still, we can construct cases with two agents, one of whom has even more and better evidence than the other, where according to contextualism attributions of knowledge to the former agent are false but attributions to the latter are true. Surely something is amiss here.

19 Feldman (1999) demonstrates this clearly.

20 See Stephen Schiffer, "Contextualist solutions to skepticism," *Proceedings of the Aristotelian Society* 96 (1996): 317–33, §IV. This is required to explain the paradoxical air of the skeptical argument, where, says the contextualist, normal speakers unschooled in contextualism do not realize that "knows" locutions sometimes change meaning depending on context.

21 For a defense of a general reliabilism that does not make belief necessary for knowledge, see Nicholas Smith, "Generic knowledge," *American Philosophical Quarterly* 39:4 (2002): 343–57.

22 Surely the contextualist will say that she is interested in *human* knowledge, which *is* context sensitive. But if what I have argued here is right, then contextualism is a superfluous addition to our understanding of *human* knowledge, too – reliabilism is enough. Contextualism nicely explains our knowledge *attributions*, but surely there is space for questioning the verity of some of those attributions, which is close to the point I express in the next sentence in the text. In "Knowing and possessing knowledge" (*American Philosophical Quarterly* 41:1 (2004): 21–36; hereafter cited as Becker (2004)), I too appeal to how ordinary reflective persons attribute knowledge possession, but I use this data only as a primer to drive a wedge between knowing and possessing knowledge – a wedge I then remove. I make no claims as to the truth of lay persons' off-hand attributions.

23 Thus suppose that the contextualist has a response to the complaint that, while ordinary knowledge attributions indicate standards of warranted assertability, they do not accurately reflect the *truth conditions* of "knows" locutions. (See, for example, DeRose (1999, §§15–16).) Extending the point in the text, one may still question whether the data to which the contextualist appeals accurately reflect standards of warranted assertability in the first place. Perhaps on sufficient reflection normal speakers would deny that their off-the-cuff knowledge attributions are even warrantedly assertable.

24 One could appeal to safety to defend the claim that I know that I'm not a BIV: In nearby worlds where *p* is true, I believe that *p*. This is the crux of Sosa's defense of a Moorean position. Ernest Sosa, "How to defeat opposition to Moore," *Philosophical Perspectives* 13 (1999): 141–53. But this appeal does not *explain* how I know that I'm not a BIV – it only "analyzes" it. The basic question remains: If I have no evidence or reliable belief-forming process that tells me I am not a BIV, how can I know it?

25 DeRose addresses this in note 36 (DeRose 1999). "I would find it embarrassing if we could never claim to have such knowledge by means of simple knowledge attributions." He directs us to his Ph.D. dissertation for the answer, but it is unclear how it would fit with the general contextualist strategy. On how knowledge "vanishes" when skeptical worries are raised, see also David Lewis, "Elusive knowledge," *Australasian Journal of Philosophy* 74:4 (1996): 549–67.

26 The value problem for reliabilism is now common coin. It is put forcefully and clearly in Zagzebski, "From reliabilism to virtue epistemology," in Guy Axtell, ed. *Knowledge, Belief, and Character* (Lanham, MD: Rowman and Littlefield, 2000), 113–22, and Zagzebski, "The search for the source of epistemic good," *Metaphilosophy* 34:1/2 (2003a): 12–28; hereafter cited as Zagzebski (2003a, "The search"). See also Sosa, "The place of truth in epistemology," in Michael DePaul and Linda Zagzebski, eds. *Intellectual Virtue* (Oxford: Oxford University Press,

2003), 155–79; hereafter cited as Sosa (2003). Zagzebski argues that the value problem afflicts any truth-conducivist theory of warrant or justification. She cites Bonjour's claim that "the basic role of justification is that of a *means* to truth" to show that the value problem cuts across the internalist-externalist divide (Zagzebski 2003a, "The search," 16). See Laurence Bonjour, *The Structure of Empirical Knowledge* (Cambridge, MA: Harvard University Press, 1985). Even an arch-internalist like Bonjour must face it if he holds that attaining truth is the only epistemic value, and that justification is merely a means to it. One might wonder whether the value problem, applied thus broadly, afflicts even Zagzebski's virtue theory. If the value of praiseworthy virtue lies in the agent's motivation to seek truth, then again it seems one could say that truth is the end that determines the goodness of an agent's character, and so true belief is all that matters. This is not right, for surely Zagzebski has located, in the agent, a good that is independent of true belief. Here I aim to exploit this general strategy, arguing that there is an independent good in the enviable strength of one's epistemic position.

27 Zagzebski's "The search" and "Intellectual motivation and the good of truth" (in Michael DePaul and Linda Zagzebski, eds. *Intellectual Virtue* (Oxford: Oxford University Press, 2003b), 135–54; hereafter cited as Zagzebski (2003b, "Intellectual motivation")) are explicitly so motivated.

28 While this is the general thrust of Zagzebski's argument (for instance, in her "The search," especially 16f.), she is clearly aware that motivational virtue may not in fact be a necessary condition for knowledge: "The value I am talking about . . . is unlikely to be the pinnacle [of intellectual virtue], but *it is probably more than the minimum we can get away with*" ("Intellectual motivation," 150 (my emphasis)). Zagzebski goes on to say that if we follow her, we can give a more satisfying explanation of the value of knowledge. One gets the impression that the topic has been changed from an explication of knowledge *per se* to an account of epistemic evaluation of the best kinds of agents. Hookway argues explicitly that the traditional project of analyzing knowledge ought to be abandoned in favor of an explanation of our practices of epistemic evaluation. Christopher Hookway, "How to be a virtue epistemologist," in Michael DePaul and Linda Zagzebski, eds. *Intellectual Virtue* (Oxford: Oxford University Press, 2003), 185–202.

29 Zagzebski notes that there is a difference for most people between mere credit and praiseworthiness: "Praiseworthiness differs from credit in most people's vocabulary, in that deserving praise is a stronger commendation than deserving credit. I think the difference is only one of degree and do not believe that much hangs on the difference" ("The search," note 24). Thus it may seem unfair to say that she claims that credit *implies* praiseworthiness. However, Zagzebski believes that an account of knowledge that sufficiently captures its unique value ought to include praiseworthiness.

30 See also Greco, "Virtue and luck, epistemic and otherwise," *Metaphilosophy* 34:3 (2003a): 353–66, Greco, "Knowledge as credit for true belief," in Michael DePaul and Linda Zagzebski, eds. *Intellectual Virtue* (Oxford: Oxford University Press, 2003b), 111–34, and Sosa (2003).

31 The notion that one must be in a strong epistemic position to know is not at all original. It is one of the central elements of DeRose's (1999) contextualism, as we saw in §A, and of precursors to that view, for instance in Unger, *Ignorance: A Case for Skepticism* (Oxford: Oxford University Press, 1975). I shall argue that strength of epistemic position can *also* explain how knowledge is more valuable than true belief. Since I wrote the original draft of my work on the value problem, I have come across Jennifer Lackey's useful paper describing cases where knowledge does not entail credit. Jennifer Lackey, "Why we don't deserve credit for everything we know," *Synthese* (in press).

32 Keith Lehrer, *Theory of Knowledge*, 2nd edition (Boulder, CO: Westview Press, 2000), 187.

33 We can ask this question, and answer it, without making any claims about necessary and sufficient conditions for knowledge. Given her favored conception of justification, an internalist might very well deny that Truetemp knows, but, for present purposes, the relevant question is solely about the added value of reliability. So again, I feel compelled to point out that the criticism under consideration here is that, if reliabilists cannot account for the value of knowledge over mere true belief, then reliabilism must be false. And again, it is enough to block this objection, then, to locate a source of value in reliable belief-forming processes. A full-dress defense of reliabilism that would satisfy an internalist is not required, which is just as well, since one does not exist.

34 The issue here is not necessarily one about the involuntariness of perceptual belief. One can always question one's beliefs, and thereby exercise control over them, but questioning one's beliefs cannot be a necessary condition for perception-based knowledge, else few would have it.

35 Again, and now for the sake of explicitness, I admit that the kind of epistemic value cited here does not constitute a *direct* reply to Zagzebski's statement of the value problem. Rather, I am questioning her presupposition that the reliabilist must locate the value in the difference between one kind of true belief and another.

Bibliography

Alston, William, "The deontological conception of epistemic justification," *Philosophical Perspectives* 2 (1988): 257–99.
——, "How to think about reliability," *Philosophical Topics* 23:1 (1995): 1–29.
——, "Perceptual knowledge," in John Greco and Ernest Sosa, eds. *The Blackwell Guide to Epistemology* (Oxford: Blackwell, 1999), 223–42.
——, *Beyond Justification* (Ithaca, NY: Cornell University Press, 2005).
Annis, David, "A contextualist theory of epistemic justification," *American Philosophical Quarterly* 15 (1978): 213–19.
Aristotle, "Nichomachean ethics," in J.L. Ackrill, ed. *A New Aristotle Reader* (Princeton, NJ: Princeton University Press, 1987).
Armstrong, David, *Belief, Truth, and Knowledge* (Cambridge: Cambridge University Press, 1973).
Axtell, Guy, "Felix Culpa: luck in ethics and epistemology," *Metaphilosophy* 34:3 (2003): 331–52.
Bealer, George, "A theory of the a priori," *Pacific Philosophical Quarterly* 81 (2000): 1–30.
Becker, Kelly, "Individualism and self-knowledge: *Tu Quoque*," *American Philosophical Quarterly* 39:3 (2002): 289–95.
——, "Knowing and possessing knowledge," *American Philosophical Quarterly* 41:1 (2004): 21–36.
——, "Is counterfactual reliabilism compatible with higher-level knowledge?" *Dialectica* 60:1 (2006): 79–84.
——, "Reliabilism and safety," *Metaphilosophy* 37:5 (2006): 691–704.
Bishop, Michael and J.D. Trout, "The pathologies of standard analytic epistemology," *Noûs* 39:4 (2005): 696–714.
Bonjour, Laurence, *The Structure of Empirical Knowledge* (Cambridge, MA: Harvard University Press, 1985).
Brown, Jessica, "The reductio argument and transmission of warrant," in Susana Nuccetelli, ed. *New Essays on Semantic Externalism and Self-Knowledge* (Cambridge, MA: MIT Press, 2003), 117–30.
——, *Anti-Individualism and Knowledge* (Cambridge, MA: MIT Press, 2004).
Brueckner, Anthony, "Unfair to Nozick," *Analysis* 51:1 (1991): 61–4.
Burge, Tyler, "Individualism and the mental," in Peter French, Theodore Uehling, Jr. and Howard Wettstein, eds. *Midwest Studies in Philosophy* 4, *Metaphysics*, 1979, 73–121.

Chase, James, "Is externalism about content inconsistent with internalism about justification?" *Australasian Journal of Philosophy* 79:2 (2001): 227–46.

Chisholm, Roderick, *Theory of Knowledge*, 2nd edition (Englewood Cliffs, NJ: Prentice-Hall, 1977).

Cohen, L. Jonathan, *An Essay on Belief and Acceptance* (Oxford: Oxford University Press, 1992).

Cohen, Stewart, "Justification and truth," *Philosophical Studies* 46:3 (1984): 279–95.

——, "Contextualism, skepticism, and the structure of reasons," *Philosophical Perspectives* 13 (1999): 57–89.

——, "Contextualism and skepticism," *Philosophical Issues* 10 (2000): 94–107.

Comesaña, Juan, "Unsafe knowledge," *Synthese* 146 (2005): 395–404.

Conee, Earl and Richard Feldman, "The generality problem for reliabilism," in Louis Pojman, ed. *The Theory of Knowledge*, 3rd edition (Belmont, CA: Wadsworth, 2003), 296–309. (Originally in *Philosophical Studies* 89:1 (1998): 1–29.)

Craig, Edward, "Nozick and the sceptic: the thumbnail version," *Analysis* 49:4 (1989): 161–2.

Davidson, Donald, "Mental events," in *Essays on Actions and Events* (Oxford: Clarendon Press, 1980), 207–25.

Davies, Martin, "Externalism and armchair knowledge," in Paul Boghossian and Christopher Peacocke, eds. *New Essays on the A Priori* (Oxford: Oxford University Press, 2000), 384–414.

——, "The problem of armchair knowledge," in Susana Nuccetelli, ed. *New Essays on Semantic Externalism and Self-Knowledge* (Cambridge, MA: MIT Press, 2003), 23–55.

DeRose, Keith, "Solving the skeptical problem," in Keith DeRose and Ted Warfield, eds. *Skepticism: A Contemporary Reader* (Oxford: Oxford University Press, 1999), 183–219. (Originally in *Philosophical Review* 104:1 (1995): 1–52.)

Dretske, Fred, "Epistemic operators," *Journal of Philosophy* 67:12 (1970): 1007–23.

——, "Conclusive reasons," *Australasian Journal of Philosophy* 49:1 (1971): 1–22.

——, "Externalism and self-knowledge," in Susana Nuccetelli, ed. *New Essays on Semantic Externalism and Self-Knowledge* (Cambridge, MA: MIT Press, 2003), 131–42.

Engel, Mylan, "Is epistemic luck compatible with knowledge?" *The Southern Journal of Philosophy* 30:2 (1992): 59–75.

Feldman, Richard, "Reliability and justification," *The Monist* 68:2 (1985): 159–74.

——, "Epistemic obligations," *Philosophical Perspectives* 2 (1988): 235–56.

——, "Contextualism and skepticism," *Philosophical Perspectives* 13 (1999): 91–114.

Foley, Richard, "Epistemic luck and the purely epistemic," *American Philosophical Quarterly* 21:2 (1984): 113–24.

Forbes, Graeme, "Nozick on scepticism," *The Philosophical Quarterly* 34:134 (1984): 43–52.

Fumerton, Richard, "Nozick's epistemology," in Steven Luper-Foy, ed. *The Possibility of Knowledge: Nozick and His Critics* (Totowa, NJ: Rowman & Littlefield, 1987), 163–81.

——, *Metaepistemology and Skepticism* (Lanham, MD: Rowman & Littlefield, 1995).

Garrett, B.J., "Nozick on knowledge," *Analysis* 43:4 (1983): 181–4.

Gettier, Edmund, "Is justified true belief knowledge?" *Analysis* 23:6 (1963): 121–3.

Goldman, Alan, "Nozick on knowledge: finding the right connection," in Steven Luper-Foy, ed. *The Possibility of Knowledge: Nozick and His Critics* (Totowa, NJ: Rowman & Littlefield, 1987), 182–96.

Goldman, Alvin, "A causal theory of knowing," *Journal of Philosophy* 64:12 (1967): 355–72.

——, "Discrimination and perceptual knowledge," *Journal of Philosophy* 73:20 (1976): 771–91.

——, "What is justified belief?" in George Pappas, ed. *Justification and Knowledge* (Dordrecht: D. Reidel, 1979), 1–23.

——, "Review of philosophical explanations," *Philosophical Review* 92:1 (1983): 81–8.

——, *Epistemology and Cognition* (Cambridge, MA: Harvard University Press, 1986).

——, "Strong and weak justification," in James Tomberlin, ed. *Philosophical Perspectives* 2 (1988): 51–69.

——, "Psychology and philosophical analysis," *Proceedings of the Aristotelian Society* 89 (1989): 195–209.

——, "Epistemic folkways and scientific epistemology," in *Liaisons: Philosophy Meets the Cognitive and Social Sciences* (Cambridge, MA: MIT Press, 1992), 155–75.

——, "A priori warrant and naturalistic epistemology," in *Pathways to Knowledge* (New York: Oxford University Press, 2002), 24–50.

Greco, John, *Putting Skeptics in Their Place* (Cambridge: Cambridge University Press, 2000).

——, "Virtue and luck, epistemic and otherwise," *Metaphilosophy* 34:3 (2003a): 353–66.

——, "Knowledge as credit for true belief," in Michael DePaul and Linda Zagzebski, eds. *Intellectual Virtue* (Oxford: Oxford University Press, 2003b), 111–34.

——, "Justification is not internal," in Matthias Steup and Ernest Sosa, eds. *Contemporary Debates in Epistemology* (Oxford: Blackwell, 2005), 257–70.

Hall, Barbara, "On epistemic luck," *The Southern Journal of Philosophy* 32 (1994): 79–94.

Harman, Gilbert, *Thought* (Princeton, NJ: Princeton University Press, 1973).

Hawthorne, John, *Knowledge and Lotteries* (Oxford: Clarendon Press, 2004).

Heil, John, "Doxastic agency," *Philosophical Studies* 43:3 (1983): 355–64.

Heller, Mark, "The simple solution to the problem of generality," *Noûs* 29:4 (1995): 501–15.

Hetherington, Stephen, *Good Knowledge, Bad Knowledge* (Oxford: Clarendon Press, 2001).

Hill, Christopher, "Process reliabilism and Cartesian scepticism," *Philosophy and Phenomenological Research* 56:3 (1996): 567–81.

Hookway, Christopher, "How to be a virtue epistemologist," in Michael DePaul and Linda Zagzebski, eds. *Intellectual Virtue* (Oxford: Oxford University Press, 2003), 185–202.

Klein, Peter, "Skepticism and closure: why the evil genius argument fails," *Philosophical Topics* 23:1 (1995): 213–36.

Kornblith, Hilary, *Knowledge and its Place in Nature* (Oxford: Clarendon Press, 2002).

Kripke, Saul, *Naming and Necessity* (Cambridge, MA: Harvard University Press, 1972).

——, *Wittgenstein on Rules and Private Language* (Cambridge, MA: Harvard University Press, 1982).

Kvanvig, Jonathan, *The Intellectual Virtues and the Life of the Mind: On the Place of Virtues in Epistemology* (Lanham, MD: Rowman and Littlefield, 1992).
——, *The Value of Knowledge and the Pursuit of Understanding* (Cambridge: Cambridge University Press, 2003).
Kyburg, Henry, *Probability and the Logic of Rational Belief* (Middletown, CT: Wesleyan University Press, 1961).
Lackey, Jennifer, "Why we don't deserve credit for everything we know," *Synthese* (in press).
Lehrer, Keith, *Theory of Knowledge*, 2nd edition (Boulder, CO: Westview Press, 2000).
Lehrer, Keith and Thomas Paxson, "Knowledge: undefeated justified true belief," *Journal of Philosophy* 66:8 (1969): 225–37.
Lewis, David, *Counterfactuals* (Oxford: Blackwell, 1973).
——, "Elusive knowledge," *Australasian Journal of Philosophy* 74:4 (1996): 549–67.
Luper-Foy, Steven, "The epistemic predicament: knowledge, Nozickian tracking, and scepticism," *Australasian Journal of Philosophy* 62:1 (1984): 26–49.
——, "The possibility of skepticism," in Steven Luper-Foy, ed. *The Possibility of Knowledge: Nozick and His Critics* (Totowa, NJ: Rowman & Littlefield, 1987), 219–41.
Lycan, William, "Evidence one does not possess," *Australasian Journal of Philosophy* 55:2 (1977): 114–26.
McGinn, Colin, "The concept of knowledge," in Peter French, Theodore Uehling, Jr. and Howard Wettstein, eds. *Midwest Studies in Philosophy* 9, *Causation and Causal Theories* (1984), 529–54.
Montmarquet, James, *Epistemic Virtue and Doxastic Responsibility* (Lanham, MD: Rowman and Littlefield, 1993).
Moore, G.E., *Philosophical Papers* (London: George Allen & Unwin, 1959).
Moser, Paul, *Knowledge and Evidence* (Cambridge: Cambridge University Press, 1989).
Neta, Ram, "Contextualism and the problem of the external world," *Philosophy and Phenomenological Research* 66:1 (2003): 1–31.
Nozick, Robert, *Philosophical Explanations* (Cambridge, MA: Harvard University Press, 1981).
Nuccetelli, Susana, ed., *New Essays on Semantic Externalism and Self-Knowledge* (Cambridge, MA: MIT Press, 2003).
Plantinga, Alvin, *Warrant and Proper Function* (New York: Oxford University Press, 1993).
Pritchard, Duncan, "Resurrecting the Moorean response to the sceptic," *International Journal of Philosophical Studies* 10:3 (2002a): 283–307.
——, "Recent work on radical skepticism," *American Philosophical Quarterly* 39:3 (2002b): 215–57.
——, "Radical scepticism, epistemological externalism, and closure," *Theoria* 68:2 (2002c): 129–61.
——, "Virtue epistemology and epistemic luck," *Metaphilosophy* 34:1/2 (2003): 106–30.
——, *Epistemic Luck* (Oxford: Oxford University Press, 2005).
Putnam, Hilary, "The meaning of 'meaning'," in *Mind, Language and Reality* (Cambridge: Cambridge University Press, 1975), 215–71.
——, *Reason, Truth and History* (Cambridge: Cambridge University Press, 1981).
Quine, W.V., *Word and Object* (Cambridge, MA: MIT Press, 1960).

——, "Ontological relativity," in *Ontological Relativity and Other Essays* (New York: Columbia University Press, 1969a), 26–68.
——, "Epistemology naturalized," in *Ontological Relativity and Other Essays* (New York: Columbia University Press, 1969b), 69–90.
Rawls, John, *A Theory of Justice* (Cambridge, MA: Harvard University Press, 1971).
Riggs, Wayne, "What are the chances of being justified?" *The Monist* 81:3 (1999): 452–72.
——, "Reliability and the value of knowledge," *Philosophy and Phenomenological Research* 64:1 (2002): 79–96.
Roush, Sherrilyn, *Tracking Truth* (Oxford: Oxford University Press, 2005).
Salerno, Joseph, "Truth-tracking and the problem of reflective self-knowledge," in Joseph Keim-Campbell and Michael O'Rourke, eds. *Topics in Contemporary Philosophy* 5: *Knowledge and Scepticism* (Cambridge, MA: MIT Press, in press).
Schaffer, Jonathan, "Contrastive knowledge," in Tamar Gendler and John Hawthorne, eds. *Oxford Studies in Epistemology: 1* (Oxford: Oxford University Press, 2005), 235–71.
Schiffer, Stephen, "Contextualist solutions to skepticism," *Proceedings of the Aristotelian Society* 96 (1996): 317–33.
Schmitt, Frederick, *Knowledge and Belief* (New York: Routledge, 1992).
Shope, Robert, "Cognitive abilities, conditionals, and knowledge: a response to Nozick," *Journal of Philosophy* 81:1 (1984): 29–48.
——, "Conditions and analyses of knowing," in Paul Moser, ed. *The Oxford Handbook of Epistemology* (Oxford: Oxford University Press, 2002), 25–70.
Smith, Nicholas, "Generic knowledge," *American Philosophical Quarterly* 39:4 (2002): 343–57.
Sosa, Ernest, *Knowledge in Perspective* (Cambridge: Cambridge University Press, 1991).
——, "Philosophical scepticism and epistemic circularity," *Aristotelian Society Supplementary Volume* 68 (1994): 263–90.
——, "How to defeat opposition to Moore," *Philosophical Perspectives* 13 (1999): 141–53.
——, "Skepticism and contextualism," *Philosophical Issues* 10 (2000): 1–18.
——, "The place of truth in epistemology," in Michael DePaul and Linda Zagzebski, eds. *Intellectual Virtue* (Oxford: Oxford University Press, 2003), 155–79.
Stalnaker, Robert, "Comments on 'From contextualism to contrastivism'," *Philosophical Studies* 119 (2004): 105–17.
Stanley, Jason, "On the linguistic basis for contextualism," *Philosophical Studies* 119 (2004): 119–46.
——, *Knowledge and Practical Interests* (Oxford: Oxford University Press, 2005).
Steup, Matthias, "A defense of internalism," in Louis Pojman, ed. *The Theory of Knowledge*, 3rd edition (Belmont, CA: Wadsworth, 2003), 310–21.
Stine, Gail, "Skepticism, relevant alternatives, and deductive closure," *Philosophical Studies* 29 (1976): 249–61.
Stroud, Barry, "Scepticism, 'externalism', and the goal of epistemology," *Aristotelian Society Supplementary Volume* 68 (1994): 291–307.
Unger, Peter, *Ignorance: A Case for Skepticism* (Oxford: Oxford University Press, 1975).
——, *Philosophical Relativity* (Oxford: Oxford University Press, 2002). (First published by University of Minnesota Press, 1984.)

Vahid, Hamid, "Knowledge and varieties of epistemic luck," *Dialectica*, 55:4 (2001): 351–62.

Vogel, Jonathan, "Tracking, closure, and inductive knowledge," in Steven Luper-Foy, ed. *The Possibility of Knowledge: Nozick and His Critics* (Totawa, NJ: Rowman & Littlefield, 1987), 197–215.

——, "Reliabilism leveled," *The Journal of Philosophy* 97:11 (2000): 602–23.

Williams, Michael, *Unnatural Doubts: Epistemological Realism and the Basis of Scepticism* (Oxford: Blackwell, 1991).

Williamson, Timothy, *Knowledge and its Limits* (New York: Oxford University Press, 2000).

Wittgenstein, Ludwig, *Philosophical Investigations* (Oxford: Blackwell, 1953).

——, *On Certainty* (Oxford: Blackwell, 1969).

Wright, Crispin, "Cogency and question-begging: some reflections on McKinsey's paradox and Putnam's proof," *Philosophical Issues* 10 (2000): 140–63.

——, "Some reflections on the acquisition of warrant by inference," in Susana Nuccetelli, ed. *New Essays on Semantic Externalism and Self-Knowledge* (Cambridge, MA: MIT Press, 2003), 57–77.

——, "Contextualism and scepticism: even-handedness, factivity and surreptitiously raising standards," *The Philosophical Quarterly* 55:219 (2005): 237–62.

Zagzebski, Linda, *Virtues of the Mind* (Cambridge: Cambridge University Press, 1996).

——, "From reliabilism to virtue epistemology," in Guy Axtell, ed. *Knowledge, Belief, and Character* (Lanham, MD: Rowman and Littlefield, 2000), 113–22.

——, "The search for the source of epistemic good," *Metaphilosophy* 34:1/2 (2003a): 12–28.

——, "Intellectual motivation and the good of truth," in Michael DePaul and Linda Zagzebski, eds. *Intellectual Virtue* (Oxford: Oxford University Press, 2003), 135–54.

Index of cases

Barn Facsimile 24
Benevolent Demon 32
Blackjack 135
Brain Lesion 32
Brenda 27

Careless Math Student 101
Chicken Sexer 13
Chicken Sexer Choosing 102
Chimney 23
Chords 89
Cleverly Disguised Mule 26
Conjunction 100

Dach 48
Dictator 46

Faithful Father 51
Fermat's Last Theorem 47
Flip 54

Gettier 14
Going Downtown 22
Goldfinch 84
Grabit 46
Grandmother 48

Halloween Party 68
Haveit 53
Hologram Vase 22

Jesse James 49

Keanu Reeves 73

Lava 22
LAX 125
Lottery 45

Mary 36

Naturally Gifted Math Student 102
Necessary Truth 45
New Evil Demon 34
Newspaper 24

Oliver Sacks 92
Omar 57

Rookie, The 54

Serendipitous Reliable Process 37
Sunrise 55
Sun Worshipper 75

Thermometer 29
Trash Chute 54
Truetemp 131
Trustworthy Parents 34

Vase-Tracking Hologram 53

Subject and name index

abominable conjunction 126
Alston, William 105, 139n2, 141n13, n18, 156–7n10, 158n31
Annis, David 144n40
argument from ignorance 69–70, 81–82, 113–18, 122–25; *see also* skeptical hypotheses, skepticism
Aristotle 96, 141n19, 157n16
Armstrong, David 139n1
Axtell, Guy 141n14, 143n30

Bealer, George 146n27
Becker, Kelly 132, 140n7, 152n28, 153n14, 154n15, 162n22
belief-forming methods (narrow) 15–16, 37, 40, 45–47, Chapter 3 (§B) *passim*, 53, 54–56, 61–62, 68, 72–77, 88–89, 94, 97–103, 106–9, 114
belief-forming processes and methods (wide) 4, 11, 12, 14–17, 19–20, 23, 25, 31, Chapter 2 (§C) *passim*, 31, 38, 40, 46, 50–51, 53–54, 61–62, 69, 71–73, 75–76, 79–80, 85–86, 88–94, 97–98, 100–110, 112, 118–21, 124, 128–34, 136
Bishop, Michael 139n5
Brown, Jessica 119, 142n26, 160n11
Brueckner, Anthony 52, 151n22
Burge, Tyler 6, 140n9, 141n19

character (intellectual) 37, 72, 75, 93, 128–29; *see also* reliabilism (agent), virtue epistemology
Chase, James 141n17
Chisholm, Roderick 140n11
closure principle 2, 18–19, 26–27, 39, 45, 52–53, 62, 63, 65, 69–70, 79–82, 84, 87, 96, 100, 110, 111–28, 137–38
Cohen, L.J. 155–6n1
Cohen, Stewart 114, 125–27, 144n40, n41, 146n20, 160n5, 162n17
Comesaña, Juan 68, 153n11
Conee, Earl 104–6, 147n33, 158n27
consequentialism 111, 128, 130–31, 136–37
contextualism 2, 18–20, 25–27, 113–18, 122–28
contrastivism 80–86
Cooney, Brian x, 142n24, 146n24, 152n27
counterfactuals 4, 10–11, 25, 28–32, 43, 48, 51–52, 55, 58, 65, 89, 96, 134; *see also* sensitivity, subjunctive conditionals
Craig, Edward 52, 151n21
credit (epistemic) 19–20, 92–94, 102–4, 107, 110, 128–30, 134–36

Davidson, Donald 76, 155n23
Davies, Martin 118, 160n10
DeRose, Keith ix, 114–17, 122–26, 140n6, 144n41, 157n14, 160n6, 162n23, n25, 163n31
Descartes, René 7, 17, 71, 73
discrimination Chapter 2 (§B) *passim*, 78– 80, 82–83, 87; *see also* distinguishing capacity, responsiveness
distinguishing capacity 10, 24–25, 29, 31, 38, 69, 78–80, 86, 89–91, 105–6, 109, 120–22; *see also* discrimination, responsiveness
Dolan, John ix
Dretske, Fred 26–28, 30, 42, 45, 145n8, n12, 146n25, 147n30, 148n4, 161n14

Engel, Mylan 142n28
entitlement 118–20
epistemic luck 11–12, 13–16, 19–20, 21–22, 24–25, 29, 32, 35–39, 40, 53–54, 61, 63–72, 75, 77, 80, 86,

92–93, 98, 108, 118, 129–30; process 15–16, 19–20, 32, 35, 36, 39, 53–54, 61, 63, 69–71, 75, 77, 80, 92–93, 108; reflective 13–14, 19, 35; veritic 13–16, 19–20, 22, 25, 35–37, 39, 40, 63–72, 86, 93, 118
epistemic position 19–20, 95, 112–13, 116, 130–32, 134–36
epistemological revisionism 69, 113–14, 117, 126
externalism: epistemological 4, 5, 7–9, 14–15, 17, 18, 20, 35, 80, 93, 119–20, 122, 129, 131–32, 137; semantic 5–9

Feldman, Richard 104–6, 147n33, 157n10, 158n27, 159n32, 161n12, 162n19
Foley, Richard 142n28
Forbes, Graeme 53, 149n17
Fumerton, Richard 100–101, 142n23, 157n22

Garrett, B.J. 47, 97, 148n13, 149n17, 157n18
generality problem 4, 34, 37, 39, 54, 88, 94, 96, 98, 103–9
Gettier, Edmund 1, 143n31
Gettier problem/counterexamples 1–2, 14–16, 21–22, 24, 34, 40, 42, 44, 53, 60
Goldberg, Sandy ix, x, 151n19
Goldman, Alan 53, 151n23, 157n21
Goldman, Alvin ix, x, 4, 7, 9, 12, Chapter 2 *passim*, 40–45, 48, 53, 78–79, 88, 95, 97, 104–6, 140n12, 142n25, n26, 144n2, n4, n5, n6, 145n11, n13, n18, 146n21, n22, n23, n26, 147n32, n33, n34, 148n5, 149n17, 155n24, 156n3, 158n29, 159n33
Goodman, Russell x
Greco, John x, 65–67, 70–72, 90, 92–94, 101–3, 129, 141n15, 142n24, 143n35, 145n18, 146n28, 153n8, 154n16, n18, 156n8, 157n11, 158n26, 163n30

Hall, Barbara 142n28
Harman, Gilbert 46, 144n5, 148n11
Hawthorne, John 124–25, 145n7, 148n7, 152n26, 153n7, 155n29
Heil, John 157n10
Heller, Mark 158n30
Hetherington, Stephen 145n14
higher-level knowledge 4, 56–60, 61–62, 63–64, 98–99
Hill, Christopher 16–17, 79, 143n36, 147n35, 155n26, 161n13
Hookway, Christopher 163n28

inductive knowledge 3, 54–56, 61–62, 63–68, 72, 98–100

Jeopardy! 141n16
justification 1, 4, 7–9, 12, 13–14, 17–18, 21, 24, 32–35, 39, 40, 93–94, 104, 112, 123, 127–28, 136; internalist 7–9, 13–14, 18, 33, 34, 39, 93–94, 112, 123, 136; responsibilist 13, 93–94, 128, 136

Klein, Peter 160n4
Kornblith, Hilary 160n8
Kripke, Saul 6, 9, 10, 140n10, 142n20, n23
Kvanvig, Jonathan 143n29
Kyburg, Henry 148n7

Lackey, Jennifer 163n31
Lehrer, Keith 131–32, 139n1, 146n23, 148n9, 164n32
Leibniz, G.W. 142n20
Lewis, David 10, 97, 142n21, 144n41, 157n17, 162n25
Luper-Foy, Steven 54, 97, 148n14, 149n17, 150n19, 157n20
Lycan, William 148n12

McGinn, Colin 78–79, 142n26, 146n16, 155n25
modalism/modalized epistemology 5, 9–12, 18–20, 21, 25, 32, 36, 40, 45–48, 61, 70–72, 75, 80, 85–87, 88, 96, 128; *see also* safety, sensitivity, reliabilism (modalized)
Montmarquet, James 143n29
Moore, G.E. 65, 114, 144n39, 153n6
Moser, Paul 146n23

naturalized epistemology 23
Neta, Ram 115, 160n7
Nozick, Robert ix, 4, 28–31, 36, Chapter 3 *passim*, 75, 88, 95, 97, 99–101, 142n23, 147n2, n3, 148n5, n8, n9, n10, n14, 149n16, n18, 150n19, 152n26, 157n15, 161n12
Nuccetelli, Susana 140n7

Paxson, Thomas 139n1, 146n23, 148n9
Plantinga, Alvin 32, 71, 87, 92, 128, 133, 145n17, 147n31, 148n8, 154n17, 156n7
Plato 1
Pritchard, Duncan x, 13–15, 36, 64–66, 69–76, 91, 122, 142n27, n28, 144n39, 145n15, 147n31, 149n16, 153n3, n4, n10, n14, n15, 154n16, n21, 155n22, n28, 156n6, 160n4, 161n14
proper function 87, 92, 132
Putnam, Hilary 5, 140n8, 148n6

Quine, W.V. 6, 23, 140n10, 144n3

Rawls, John 94, 157n12
relevant alternatives Chapter 2 (§B) *passim*, 38, 41–44, 83–85, 94–96, 122
reliabilism (process) 4, 11–12, 15–19, Chapter 2 *passim*, 40, 45–48, 50–51, 53–54, 61–62, 63, 69–73, 75–80, 86–87, 89–98, 101–10, 111–12, 117–24, 126–38; actual world 32–34, 71–72, 79, 121, 128; agent 36–37, 69, 71–72, 77, 87, 92, 101, 103–4, 107, 129–30; global 12, Chapter 2 *passim*, 53–54, 78; local 12, 21, 31, 37, 39, 40, 78, 86; modalized 12, 36, 40, 45–48, 61, 70, 72, 75, 87, 128
responsiveness 19, 40, 62, 69, 77–80, 87, 92, 121–22; *see also* discrimination, distinguishing capacity
Riggs, Wayne x, 112, 129–30, 134, 136, 142n28, 159n1
Roush, Sherrilyn 78, 153n2

safety 2, 12, 18, 36, 62, Chapter 4 *passim*, 89, 96, 110, 134
Salerno, Joseph 152n29
Schaffer, Jonathan 80–81, 85, 155n29, n31, n32, n33
Schiffer, Stephen 162n20
Schmitt, Frederick 158n30
Schueler, Fred x
sensitivity 4–5, 11–12, 15–16, 18, 36, 39, Chapter 3 *passim*, 63–65, 67–70, 75–80, 86–87, 88, 94–101, 106–10, 111, 113–14, 116–18, 120–24, 134; *see also* counterfactuals, subjunctive conditionals
Shope, Robert 49–50, 55, 109, 149n17, 151n24, 159n34
skeptical hypotheses 2, 16–17, 25, 27, 38, 45, 52, 60, 65, 69–70, 75–80, 82, 87, 111, 115–25, 127
skepticism 6–7, 9, 12–14, 16–17, 18, 21, 34, 38, 44–45, 52, 55, 60, 65, 69–70, 81–82, 85, 87, 111, 113–20, 124–28, 137; *see also* skeptical hypotheses
Smith, Nicholas 162n21
Sosa, Ernest 17, 36–37, 54, 56, 63–65, 67, 72–74, 99, 129, 143n37, 146n26, n28, 147n31, 152n25, 153n1, n2, n13, 154n18, 162n24, n26, 163n30
Stalnaker, Robert 155n31
Stanley, Jason 2, 139n3, 145n7
Steup, Matthias 8, 140n11, 157n10
Stine, Gail 27, 145n9
Stroud, Barry 17, 143n38, n39
subjunctive conditionals 10–11, 28–31, 36, 39, 41–49, 51, 58, 60–61, 75, 95–97; "pure" subjunctive conditional 28–30, 44, 95–96; *see also* counterfactuals, sensitivity

Taber, John x
Thomson, Iain x
Trout, J.D. 139n5
twin earth 5–6

Unger, Peter 159n2, 163n31

Vahid, Hamid 142n28
value problem 19–20, 87, 94, 110, 112, Chapter 6 (§B) *passim*
virtue epistemology 18–20, 36–37, 69–71, 77, 87, 92–93, 101, 103, 110, 128–30, 134, 136; *see also* character (intellectual), reliabilism (agent)
Vogel, Jonathan 48, 54, 56–60, 62, 98, 142n26, 149n15, 152n26, n28, n29, 153n30

Warfield, Ted ix
warrant 8, 14, 19, 35, 66, 80, 116, 118–20, 122–23, 127; transmission 118–19
Williams, Michael 160n8
Williamson, Timothy 150n19
Wittgenstein, Ludwig 6, 140n10, 144n39
Wright, Crispin 118–20, 123, 145n7, 160n10

Zagzebski, Linda 14, 19, 32, 36, 66, 72, 129, 131–32, 136, 143n29, 144n42, 147n29, n31, 153n9, 156n9, 162n26, 163n27, n28, n29, n30, 164n35

For Product Safety Concerns and Information please contact our EU representative GPSR@taylorandfrancis.com
Taylor & Francis Verlag GmbH, Kaufingerstraße 24, 80331 München, Germany

www.ingramcontent.com/pod-product-compliance
Ingram Content Group UK Ltd.
Pitfield, Milton Keynes, MK11 3LW, UK
UKHW022125190926
475702UK00013B/1648

* 9 7 8 0 4 1 5 5 4 1 2 7 5 *